Pro/ENGINEER Quick Reference

Everything you want to know about Pro/ENGINEER, fast!

By the OnWord Press Development Team

Pro/ENGINEER Quick Reference

Everything you want to know about Pro/ENGINEER, fast!

By the OnWord Press Development Team

Published by:

OnWord Press
1580 Center Drive
Santa Fe NM 87505 USA

10 9 8 7 6 5 4 3 2 1

Printed in the United States of America

Library of Congress Cataloging-in-Publication Data

OnWord Press Development Team
Pro/ENGINEER Quick Reference

Includes index.

1.Pro/ENGINEER (computer program) I. Title

93-84598

ISBN 1-56690-036-0

Trademarks

Warning and Disclaimer

About the Authors

The OnWord Press Development Team is a group of skilled editors and writers who have created a number of successful books about CAD and Unix applications.

Special Thanks

The OnWord Press Development Team wishes to extend special thanks to Amanda Radice and Lou Volpe of Parametric Technology Corporation for their cooperation and support in furthering this project. Special thanks also to Steve Cruickshank for always being available with answers to our technical questions.

And special thanks to Debbie Chen at Silicon Graphics Corporation for her endless efforts to provide us with the necessary hardware for this project.

Book Production

This book was produced using Ventura 4.1 Desktop Publishing software running on 80386 and 80486 PCs. Text was produced with a variety of word processing programs. The cover was produced by Lynne Eigensteiner with Photoshop 2.5 and Quark 3.1 on a Mac Quadra 700.

OnWord Press

OnWord Press is dedicated to the fine art of practical software user's documentation.

In addition to the authors who developed the material for this book, other members of the OnWord Press team make the book end up in your hands.

Thanks to Terry Cline, Dan Raker, David Talbott, Michelle Emmons, Margaret Burns, Carol Leyba, Lynne Eigensteiner, and the numerous other members of the OnWord Press team who contributed to the production and distribution of this book.

Table of Contents

Introduction

The *Pro / ENGINEER Quick Reference* is the fastest way to find answers to your questions about Pro/ENGINEER's commands and command structure. With commands grouped by design mode, menu listing and indexed alphabetically, you can quickly find the information you need.

This quick reference covers the commands included in the basic Pro/ENGINEER software package, which includes the following modules:

- ProFEATURE
- ProINTERFACE
- ProDETAIL
- ProDESIGN
- ProASSEMBLY
- ProSURFACE
- ProPLOT

NOTE: If you have additional design modules, some of the menus shown in this book may differ from the ones you see on your screen. Additional design modules add more menu selections and more submenus.

How This Book Is Organized

The *Pro / ENGINEER Quick Reference* is divided into six sections:

- Common Commands
- Datums
- Sketcher
- Part
- Assembly
- Drawing

Except for Common Commands, these sections correspond to important design functions and operational modes. The Common Commands section includes commands that appear on many menus. It also contains commands that affect the overall design system.

Within each section, command submenus appear essentially in the alphabetical sequence of the command that opens them. Commands for each menu are listed alphabetically.

The index at the end of the book provides an alphabetical listing of all the commands, menu names and major command categories.

Page Organization and Typographical Conventions

Finger tabs on the right hand pages and words at the very top of the left hand pages tell you at a glance which of the six sections you are in. The word at the top of the right hand page tells the general category of commands within that section.

The bold italics type above the menus shows the exact sequence of menu picks needed to get to the commands being defined. These menu picks usually begin at the MODE menu.

The menus themselves are usually the last two, three or four menus in the menu hierarchy. Bold reverse type in the menus show which items to select to open the next level of submenu. The commands on the final menu are the ones listed alphabetically and described.

NOTE: Any information of special note will appear in a note like this one.

Pro/ENGINEER Quick Reference

Everything you want to know about Pro/ENGINEER, fast!

By the OnWord Press Development Team

Section One

COMMON COMMANDS

Confirmation Commands

Accept

This command works with the Query Select command to identify hidden features for processing. It confirms the currently highlighted items as the geometry to be processed.

Done

This command is active in many Pro/ENGINEER sub-menu locations. However, the nature of the command is common to all instances. It informs Pro/ENGINEER that you have completed the current task and are satisfied with the results. Pro/ENGINEER proceeds with the process.

Next

This command works with the Query Select command to identify hidden features for processing. It steps through the next possible geometry choice and highlights it in red. When no more choices exist, Next quits functioning.

Previous

This command works with the Query Select command to identify hidden features for processing. It returns to the previous highlighted geometry choice and highlights it in red.

Quit

This command is active in many Pro/ENGINEER locations. However, the nature of the command is common to all instances. It informs Pro/ENGINEER that you have completed the current task and are not satisfied with the results.Pro/ENGINEER abandons the process.

Environment

ENVIRONMENT

MAIN
Mode
Project
Dbms
Environment
Misc
Exit
Quit Window
ChangeWindow
View

ENVIRONMENT
Tol ON
Tol OFF
Bell ON
Bell OFF
Datums ON
Datums OFF
Axes ON
Axes OFF
Wireframe
Hidden line
No hidden
Disp Tan
No Disp Tan
Tan Ctrln
Grd Snp ON
Grd Snp OFF
Colors ON
Colors OFF
Isometric
Trimetric
Thick Cables
Center Line
Csys ON
Csys OFF
Store Disp
No Display
Done / Return

Axes OFF

Turns **off** the display of datum Isometric axes.

Axes ON

Turns **on** the display of datum axes.

Bell OFF

Turns **off** the audible warning/prompt bell.

Bell ON

Turns **on** the audible warning/prompt bell.

Center Line

Makes cable elements display as center lines.

Colors OFF

Turns **off** the display of any user defined colors.

Colors ON

Turns **on** the display of any user defined colors.

Csys OFF

Turns **off** the display of coordinate systems.

Csys ON

Turns **on** the display of coordinate systems.

Datums OFF

Turns **off** the display of datum planes, datum points, and datum axes names. Datum curves are unaffected by this command.

Datums ON

Turns **on** the display of datum planes, datum points, and datum axes names. Datum curves are unaffected by this command.

Disp Tan

Turns **on** the display of an imaginary line at the intersection of two tangent surfaces. Using this command, the line is displayed as a solid line.

Grd Snp OFF

Turns **off** the ability to force all data points to a displayed grid. Most commonly used in sketcher.

Grd Snp ON

Turns **on** the ability to force all data points to a displayed grid. Most commonly used in sketcher.

Hidden Line

Makes Pro/ENGINEER models display visible lines in white and hidden lines in grey.

Isometric

Works in conjunction with the VIEW ➥ DEFAULT option to establish a model's display orientation.

No Disp Tan

Turns **off** the display of any lines at the intersection of two tangent surfaces.

No Display

Works with the file STORE and RETRIEVE commands. When selected, elements are not stored with their most recent screen display. This is the default setting.

No Hidden

Makes Pro/ENGINEER models display visible lines in white; no hidden lines are displayed.

Store Disp.

Works with the file STORE and RETRIEVE commands. When selected, elements are stored with their most recent screen display. This makes the retrieval of objects process faster, since the graphics are not recalculated.

Tan Ctrln

Turns **on** the display of an imaginary line at the intersection of two tangent surfaces. Using this command, the line is displayed as a center line.

Thick Cables

Makes cable elements display as 3D tubes. In this display mode you may shade the cables.

Tol. OFF

Turns **off** the display of dimensional tolerances.

Tol. ON

Turns **on** the display of dimensional tolerances.

Trimetric

Works in conjunction with the VIEW ➥ DEFAULT option to establish a model's display orientation.

User Def

Works in conjunction with the VIEW ➡ DEFAULT option to establish a model's display orientation. This option is only available when the values **x_angle** and **y_angle** have been set in the CONFIG.PRO file.

Wireframe

Makes Pro/ENGINEER models display in all white lines. Fastest method, but difficult to orient the model.

File And Object Management

MAIN
Mode
Project
Dbms
Environment
Misc
Exit
Quit Window
ChangeWindow
View

DBMS
Store
Path-Store
Copy
Rename
Erase
Purge
DeleteAll
Done-Return

NOTE: Some of the commands described below apear on menus other than the ones shown above.

Change Window

This option is used to move from an active window to an inactive window.

Copy

Makes an exact duplicate of an object under a new object name. If used to copy an assembly, you have the option of also copying all the components and sub-assemblies under new names.

Create

Used to create a new object. The type of object depends upon the current mode (part, assembly, drawing).

Delete All

Removes (from memory *and* from the hard disk) the currently active object. This command will not function until assemblies or drawings that reference the object are also deleted.

Erase

Removes (from memory) the currently active object. **Erase** has no effect on files located on the hard disk. This command will not function while assemblies or drawings which reference the object are still active. If used on an assembly, **Erase** does not erase the component parts.

Exit

This command ends your current Pro/ENGINEER design session. (No work is saved automatically.)

Inst — Index

Creates or updates an **instance index file** containing the name(s) of a family table "generic" and the instances in memory.

List

This option displays an information window with a listing of all object names of the current type. The objects displayed are in the current directory or exist in memory.

Purge

Pro/ENGINEER maintains separate versions of objects based upon their revision numbers. This command removes all versions of a specified object except the most current. This command will only work on objects in the current directory.

Quit Window

This option is used to close an active window. If the window is not the main window, it will be removed from the screen. If the window is the main window, it will be blanked, and you may continue designing in this window.

Rename

Changes an object name to a new definition. This command may cause problems when the object renamed is part of an assembly.

Retreive

Used to retrieve an existing object. The type of object depends upon the current mode (part, assembly, drawing).

Search / Retreive

Used to retrieve an existing object via a scrollable namelist which displays sub-directories and object names, of the current type. The objects displayed are in the current directory or exist in memory.

Store

Saves any work done in the currently active model, assembly or drawing in the current directory. You will be prompted for a filename. Do not include the extension as it will be determined by Pro/ENGINEER.

Store — Path

Saves any work done in the currently active model, assembly or drawing in a user specified directory. You will be prompted for a directory, and the filename. Do not include the extension as it will be determined by Pro/ENGINEER.

System Commands

MISC

MAIN
Mode
Project
Dbms
Environment
Misc
Exit
Quit Window
ChangeWindow
View

MISC
List Dir
Show Dir
Change Dir
System
Load Config
Edit Config
Trail
Train
SystemColors
List Options
Product Info
Picture
Time

Change Dir

Changes the current directory assignment to a new location.

Edit Config.Pro

Initiates the Pro/TABLE module to interactively modify the CONFIG.PRO configuration setting file.

List Dir

Lists the contents of any specified directory. The current directory is the default.

List Options

Lists all the optional Pro/ENGINEER modules available on your system.

Load Config.Pro

Initializes any settings found in the CONFIG.PRO file. This option must be used immediately following the **Edit Config.Pro** command for any modifications to become active.

Product Info

Displays the current product version, the **build number** for your license.

Show Dir

Displays the name of the current directory.

System

Directly accesses the workstation's operating system (without exiting Pro/ENGINEER). DOS users will recognize this command as a **shell** function. To exit this environment and return to Pro/ENGINEER, enter **Exit**.

System Colors

Pro/ENGINEER uses an established set of color assignments for various meanings. This option modifies the default system color assignments.

Time

Displays the current time in the message window.

Trail

This command is used to replay a trail file. Trail files are automatically created by Pro/ENGINEER and contain a record of all input for a working session. Trail files must be renamed and modified prior to using this command. Trail files are most often used to recover from an aborted design session, or to replay design tasks for demonstration purposes.

Train

This command is used to replay a Pro/ENGINEER or user created **training** file. Training files are a combination of session trail files and related comments. The comments will be displayed in an information window during the trail file's execution.

MISC ➡ SYSTEM COLORS

MAIN
Mode
Project
Dbms
Environment
Misc
Exit
Quit Window
ChangeWindow
View

MISC
List Dir
Show Dir
Change Dir
System
Load Config
Edit Config
Trail
Train
SystemColors
List Options
Product Info
Picture
Time

SYSTEMCOLORS
Change
Default
InitConfig
Restore
Alternate

Alternate

Pro/ENGINEER has three delivered system color palettes in addition to the default. This command toggles to the next available color palette. After the third alternate is toggled, the system returns to the first. Choose **Default** to return to the default system color palette.

Default

Changes the system color assignments back to Pro/ENGINEER'S default palette.

Init Config

Similar to the default option, but this command returns the system color assignments to the one found in the CONFIG.PRO file.

Restore

Retrieves an .scl file via a menu driven search file. This ".scl" file contains system color assignments, which will be activated upon retrieval.

SELECTION TOOLS

Done Sel

Accepts the picks made, and continues processing the current command.

ID

Selects an object (feature, part, sub-assembly) based upon an internal database number. This number remains constant even if the feature is reordered or suppressed. Use the **Model Info** command to display a feature's internal ID.

Name

Selects an object (feature, part, sub-assembly) based upon a given name. The types of names vary from **user defined** names, given at the time of object creation, to system given names like datum points and axes.

Number

Selects an object (feature, part, sub-assembly) based upon its number in the regeneration sequence.

Pick

The default selection method. Used to select visible geometry and all datums.

Pick Box

Selects all geometry that falls within a variable (rubberbanding) rectangle.

Pick Chain

Selects all geometry that is connected end to end in a continuous chain.

Pick Many

A selection method available only in drawing mode. This command uses various options to select many objects at one time.

Pick Poly

Selects all geometry that falls within a user defined area consisting of many points.

Query Select

A selection method used to select hidden geometry. Also used to select visible geometry in congested areas.

Quit Sel

Informs Pro/ENGINEER not to accept the picks, and quits the current process.

Sel by Menu

A selection method using object names as the criteria.

Unsel Last

Allows you to unselect the last selected item. This command is recursive, allowing you to continue unselecting items in the reverse order that they were picked.

Unsel Item

Allows you to selectively unselect any item you have previously picked. This option can be used with the **Pick Box** command.

Views

VIEW

MAIN
Mode
Project
Dbms
Environment
Misc
Exit
Quit Window
ChangeWindow
View

MAIN VIEW
Alter View
Names
Repaint
Orientation
Pan / Zoom
Cosmetic
Regen View
Perspective
Done / Return

Alter View

Available in assembly mode only. This command allows you to switch windows while placing components in an assembly.

Cosmetic

Many options to modify the appearance of objects (shaded views, meshed views) are available under this command.

Names

Used to create, delete, and recall saved views by user defined names.

Orientation

Used to change a view's camera position. Many options exist to rotate a view rather than an object, to enhance visual clarity.

Pan / Zoom

Used to modify a view's scale or positioning. Most often used to get a closer look at a feature, or to step away and look at the big picture.

Perspective

Changes the current display to a "one-point" perspective view, to enhance the realism of the object being displayed. Requires you to enter a focal-length value between 1 and 10,000. The smaller the value, the more distorted the perspective will be.

Regen View

Used to recalculate an object's geometry and update the current display.

Repaint

Used to update the current display without regenerating the object's geometry.

Views - Cosmetic

VIEW ➡ COSMETIC

MAIN
Mode
Project
Dbms
Environment
Misc
Exit
Quit Window
ChangeWindow
View

MAIN VIEW
Alter View
Names
Repaint
Orientation
Pan / Zoom
Cosmetic
Regen View
Perspective
Done / Return

COSM VIEW
Shade
Mesh Surf
Explode
Un-explode
Colors
Lights
Lines

Colors

This command has several options for creating user defined colors, then assigning those colors to objects. This command can be used in assembly mode, to assign unique colors to the various sub-assemblies and components.

Explode

Available only in assembly mode. This command automatically creates an exploded view of an assembly. The components are not physically moved; the view is just altered to appear that they were.

Lights

This command allows you to create new light sources or modify existing lights to enhance the results from shading your model. This command also allows you to modify the surface properties of elements. (Surface properties are established during the color definition process, and determine how an object reflects light.)

Lines

This option is only available on Silicon Graphics workstations. When available, this option enables anti-aliasing capabilities to visually smooth non-vertical lines.

Mesh Surf

Applies a mesh pattern to a specified surface, to enhance viewing clarity.

Shade

This command works with several optional settings to shade an object for increased realism.

Un — Explode

Available only in assembly mode. This command removes the effects of the **Explode** command.

VIEW ➥ COSMETIC ➥ LIGHTS

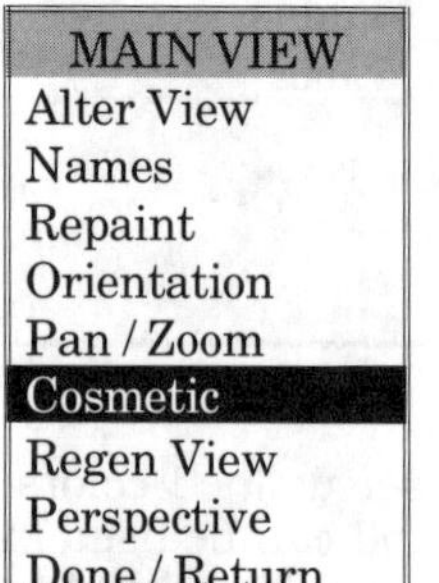

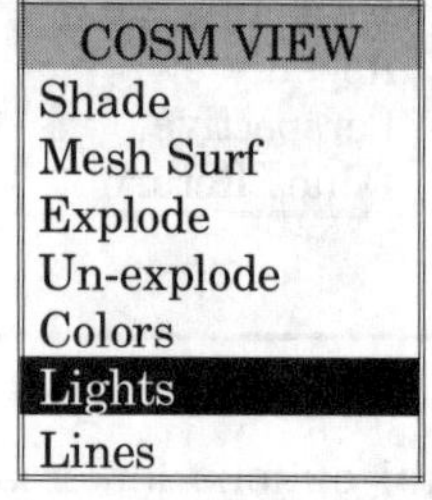

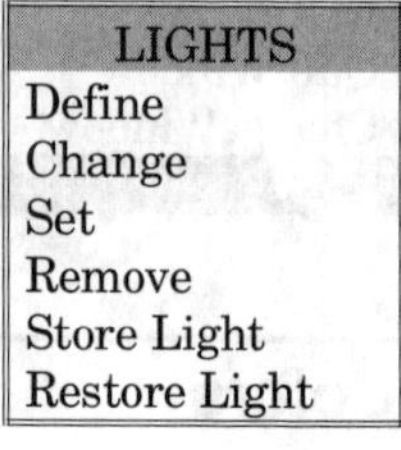

Define

Used to create new light souce definitions.

Remove

Used to erase unneccessary light sources.

Restore Light

Retrieves light definitions saved in a **light file** from an earlier Pro/ENGINEER design session.

Set

This command is used to specify which light sources are on or off.

Store Light

Stores user defined light definitions in a **light file** on the hard disk. The default file name is LIGHTS.LGH.

VIEW ➥ COSMETIC ➥ LIGHTS ➥ DEFINE

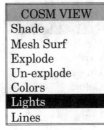

MAIN VIEW
Alter View
Names
Repaint
Orientation
Pan / Zoom
Cosmetic
Regen View
Perspective
Done / Return

COSM VIEW
Shade
Mesh Surf
Explode
Un-explode
Colors
Lights
Lines

LIGHTS
Define
Change
Set
Remove
Store Light
Restore Light

Ambient Light
Direction Light
Point Light
Spot Light

Ambient Light

Used to define the ambient light properties around an object. This light type effects all surfaces equally.

Direction Light

Used to define a light source at an infinite distance from an object (like sunlight) emanating from a specific direction. All the rays of this light type are parallel.

Point Light

Used to define a light source at a specified distance from an object emanating from a given point (like a light bulb).

Spot Light

Used to define a light source at a specified distance from an object. This light type has a controllable light spread, defined by the size of the light's cone (like a flashlight).

VIEW ➡ COSMETIC ➡ SHADE

MAIN VIEW
Alter View
Names
Repaint
Orientation
Pan / Zoom
Cosmetic
Regen View
Perspective
Done / Return

COSM VIEW
Shade
Mesh Surf
Explode
Un-explode
Colors
Lights
Lines

SHADE
Display
Save
Restore
Quality
To PostScript

Display

Instructs Pro/ENGINEER to shade an object as it is displayed in the window (using the current quality setting).

Quality

Specifies a value between one and ten (ten being the highest quality, three being the default setting), that controls the size of triangles used to shade revolved surfaces. The higher the quality setting, the smoother rounded surfaces will appear.

Save

Saves the image of a shaded object to a ".shd" file on the hard disk. The name of the image is the same as the object name unless otherwise specified.

To Postscript

Saves the image of a shaded object to an Encapsulated PostScript (.eps) file on the hard disk.

VIEW ➡ COSMETIC ➡ MESH SURF

MAIN VIEW
Alter View
Names
Repaint
Orientation
Pan / Zoom
Cosmetic
Regen View
Perspective
Done / Return

COSM VIEW
Shade
Mesh Surf
Explode
Un-explode
Colors
Lights
Lines

MESH
Chng Mesh
Selectsrf

Chng Mesh

Changes the density of a mesh pattern. You will be prompted for the number of vertical "U-mesh" and horizontal "V-mesh" lines.

NOTE: Meshed surface patterns are only visible until the view is repainted.

VIEW ➡ COSMETIC ➡ COLORS

MAIN VIEW
Alter View
Names
Repaint
Orientation
Pan / Zoom
Cosmetic
Regen View
Perspective
Done / Return

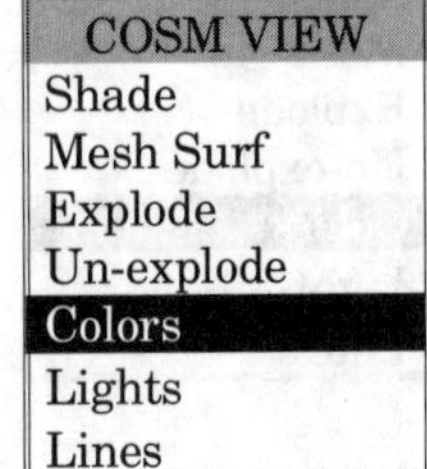

COSM VIEW
Shade
Mesh Surf
Explode
Un-explode
Colors
Lights
Lines

COLORS
Show
Define
Set
Unset
Change
Store Map
Done

Change

Allows you to modify an existing color definition (mixture of red, green, and blue values). The changes are not automatically stored; you must use the **Store Map** option to save your changes.

Define

Allows you to create a new color definition (mixture of red, green, and blue values) to use with objects.

Set

Used to assign colors to parts, assemblies, or sub-assemblies. You may also use the Set ➡ SelSurfaces option to assign a color to a part's surfaces for use with shaded images.

Show

Displays any user defined color definitions.

Store Map

Saves any new color definitions, changes to existing definitions or assignments made during the current design session. The settings are stored in a file named COLOR.MAP.

Unset

Used to remove any color assignments given to an object (part, assembly, sub-assembly, or surface).

VIEW ➡ COSMETIC ➡ COLORS ➡ CHANGE

MAIN VIEW
- Alter View
- Names
- Repaint
- Orientation
- Pan / Zoom
- Cosmetic
- Regen View
- Perspective
- Done / Return

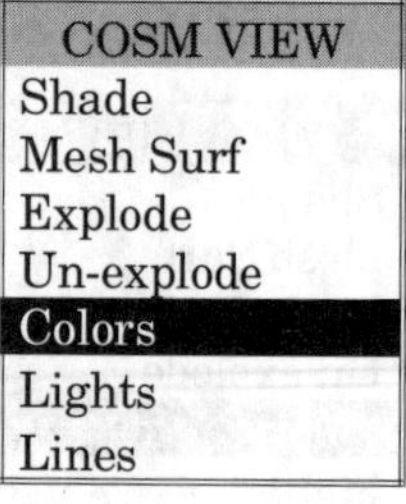

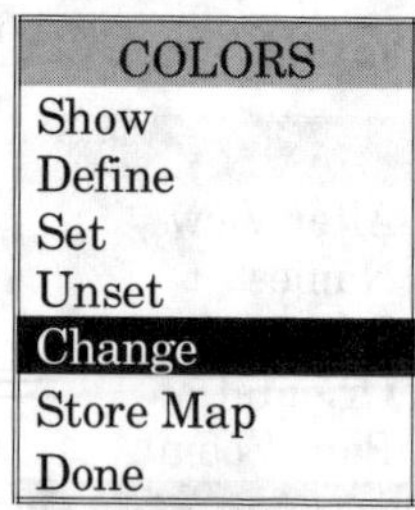

CHANGE
- Palette
- Model

Model

Modifies the color assigned to a feature. Does not affect the color definition in the current color palette.

Palette

Modifies the color definition in the current color palette. Does not affect any features that have this color assignment.

VIEW ➥ COSMETIC ➥ SHADE ➥ TO POSTSCRIPT

MAIN VIEW
Alter View
Names
Repaint
Orientation
Pan / Zoom
Cosmetic
Regen View
Perspective
Done / Return

COSM VIEW
Shade
Mesh Surf
Explode
Un-explode
Colors
Lights
Lines

Printer Type

Used to select a specific PostScript printer type. If your printer is not shown, try each of the supplied options.

Resolution

Used to specify an output resolution for the printer (100, 200, 300 or 400 dpi). The default value is **100 dpi**.

Image Size

Allows you to choose a standard image size, or specify a custom size using inches or millimeters.

Output to PS

Instructs Pro/ENGINEER to output a file to FILENAME.PLT where filename is the name you specify.

> *NOTE: Postscript files by default are set to* **portrait** *mode. To change this setting, modify the CONFIG.PRO file setting* **Landscape_Plotting**.

Views - Names

VIEW ➥ NAMES

MAIN
Mode
Project
Dbms
Environment
Misc
Exit
Quit Window
ChangeWindow
View

MAIN VIEW
Alter View
Names
Repaint
Orientation
Pan / Zoom
Cosmetic
Regen View
Perspective
Done / Return

SAVE / RETR
Retrieve
Save
Delete Name

Delete Name

Deletes a saved view from the current object.

Retrieve

Retrieves and re-displays a named view.

Save

Saves the current view settings under a user defined name. The name is limited to 80 characters.

> *NOTE: The saved view is based upon the* **default** *view assignment in the CONFIG.PRO file. Changing this assignment will have an impact on any saved views.*

Views - Orientation

VIEWS ➡ ORIENTATION

MAIN	MAIN VIEW	ORIENTATION
Mode	Alter View	Default
Project	Names	Front
Dbms	Repaint	Back
Environment	**Orientation**	Top
Misc	Pan / Zoom	Bottom
Exit	Cosmetic	Left
Quit Window	Regen View	Right
ChangeWindow	Perspective	Angles
View	Done / Return	Axis-Vert
		Axis-horiz
		Spin

Angles

Used to reorient the display of an object by specifying a degree of rotation about a selected axis.

Axis — Horiz

Used to reorient the display of an object, where a user specified axis defines the horizontal rotation axis.

Axis — Vert

Used to reorient the display of an object, where a user specified axis defines the vertical rotation axis.

Back

Defines a selected surface or edge as the **back** of a model. The selected element is one of two required.

Bottom

Defines a selected surface or edge as the **bottom** of a model. The selected element is one of two required.

Default

A system defined view orientation designed to show all three axes of an object simultaneously.

Front

Defines a selected surface or edge as the **front** of a model. The selected element is one of two required.

Left

Defines a selected surface or edge as the **left** of a model. The selected element is one of two required.

Right

Defines a selected surface or edge as the **right** of a model. The selected element is one of two required.

Spin

Dynamically reorients an object by the center of a model, by a specified edge / axis, or by using the thermometer scale.

Top

Defines a selected surface or edge as the **top** of a model. The selected element is one of two required.

VIEW ➡ ORIENTATION ➡ ANGLES

MAIN VIEW
Alter View
Names
Repaint
Orientation
Pan / Zoom
Cosmetic
Regen View
Perspective
Done / Return

ORIENTATION
Default
Front
Back
Top
Bottom
Left
Right
Angles
Axis-Vert
Axis-horiz
Spin

ANGLES
Horiz
Vert
Norm
Edge / Axis
Done / Accept
Quit / Abort

Edge / Axis

Reorients the display clockwise about a selected edge or datum axis. Using this method, the edge will act as an imaginary axis .

Horiz

Reorients the display about an imaginary axis running from left to right across the bottom of the screen.

Norm

Reorients the display about an imaginary axis running out of the screen.

Vert

Reorients the display about an imaginary axis running from bottom to top along the left edge of the screen.

VIEW ➡ ORIENTATION ➡ SPIN

MAIN VIEW
Alter View
Names
Repaint
Orientation
Pan / Zoom
Cosmetic
Regen View
Perspective
Done / Return

ORIENTATION
Default
Front
Back
Top
Bottom
Left
Right
Angles
Axis-Vert
Axis-horiz
Spin

SPIN
Center
Edge / Axis
CoordinateSys

Center

Spins a model while leaving the center of the model in place.

Edge / Axis

Spin the model about a selected edge or axis.

Views - Pan and Zoom

VIEW ➡ PAN / ZOOM

MAIN VIEW
Alter View
Names
Repaint
Orientation
Pan / Zoom
Cosmetic
Regen View
Perspective
Done / Return

PAN-ZOOM
Pan
Zoom In
Zoom Out
Reset
Magnify

Magnify

Used to identify a part of a model to view more closely. Unlike the **Zoom In** command, this option opens a new window where the magnified portion of the model will be displayed.

Pan

Used to select a new point, which will then be moved to the center of the screen when the view is re-displayed.

Reset

Resizes the display so an entire model is displayed on the screen. This command does so while maintaining the current view orientation.

Zoom In

Used to identify a part of a model to view more closely.

Zoom Out

Used to move the camera further away from an object, thereby making the object appear smaller.

Section Two

SKETCHER

Sketcher Commands

MODE ➡ SKETCHER

MAIN	MODE	SKETCHER
Mode	**Sketcher**	Restart
Project	Part	Sketch
Dbms	Sheet Metal	Pick Curve
Environment	Composite	Tools
Misc	Assembly	Modify
Exit	Drawing	Regenerate
Quit Window	Manufacture	Delete
ChangeWindow	Mold	Undelete
View	Layout	Dimension
	Format	Align
	Report	Unalign
	Markup	Interface
	Diagram	Set Up
		Toggle
		Relation
		Done
		Quit

Align

This command is used to associate the endpoints of sketched elements with existing part geometry. You may also use this command to align the centers of arcs and circle elements.

Arcs

Tangent arcs may be placed in sketcher mode by selecting a point on the endpoint of an existing element with the **right** mouse button. The arc placement command is terminated with the **middle** mouse button.

Circles

Circles may be placed in sketcher mode by selecting center and radius points with the **middle** mouse button. The circle placement command is terminated with the **left** mouse button.

Delete

This command erases unwanted sketched elements from a sketcher section. The elements are not actually deleted from memory, but are instead stored in a temporary buffer until the sketch is regenerated. You may *undelete* elements (in the order in which they were deleted) until the next time you run the **Regenerate** command.

Dimensions

Dimensions are a critical part of Pro/ENGINEER, and their use extends far beyond sketching sections. The dimensions you create will be used to *regenerate* the sketch. They will also control the feature throughout the life of the project, so be sure to select the dimensioning scheme which captures the feature's design intent, and allows the most flexibility. There are several dimension types from which to choose. (See SKETCHER ➥ DIMENSION — Types.)

NOTE: Sections may be redefined to change the dimension scheme at any time.

Interface

This command takes you to the INTERFACE menu, for file import / export capabilities.

Lines

Lines may be placed in sketcher mode by selecting beginning and ending points with the **left** mouse button. The line placement command is terminated with the **middle** mouse button.

Modify

This command allows for the manipulation of various element types. The function varies according to the element type selected. It is primarily used to change dimension values.

Regenerate

This command solves a sketched section, based upon the dimensioning scheme you created. If the section is over- or under-constrained, you will prompted by visual clues. (See Chapter Three of *Inside Pro/ENGINEER* for sketcher **Hints and Tips**.)

Relation

This command allows you to create relationships between sketched elements. The functionalities of the sketcher relations are the same as all others. (See **Relations** in the PART section of this book.

Restart

Use this command when you have already created sketched geometry, but have decided you would like to erase it and begin again. At the Sketcher initialization menu, choose create, retrieve, search and retrieve, or another option.

Setup

This option provides access to the SETUP sub-menu. The commands on the SETUP sub-menu are used to establish, dimensional attributes (accuracy, number of decimals).

Sketch

This option provides access to the GEOMETRY sub-menu. The commands on the GEOMETRY sub-menu are used to create special sketched elements.

Tools

This option provides access to the TOOLS sub-menu. The commands on the TOOLS sub-menu are used to manipulate existing sketched elements. The TOOLS sub-menu also provides access to grid modification commands.

Unalign

This command is used to remove the association of sketched element endpoints with existing part geometry. This association must have been previously created using the **Align** command.

Undelete

This command is used to restore an element that has been previously deleted and is still in the holding buffer.

Dimensions

SKETCHER ➡ DIMENSION

SKETCHER
Restart
Sketch
Pick Curve
Tools
Modify
Regenerate
Delete
Undelete
Dimension
Align
Unalign
Interface
Set Up
Toggle
Relation
Done
Quit

DIMENSION
Normal
Reference
Known
Baseline

Baseline

Used to create a baseline dimension that will be used in the creation of an ordinate dimensioning scheme. These dimensions eventually become regular part dimensions.

Known

Dimensions that reference **only part geometry**. They are used to drive sketcher dimensions through the establishment of relations. These dimensions are not displayed anywhere except in the section.

Normal

This is the default type of dimension, used to establish dimensions between sketched elements or between a sketched element and part geometry.

Reference

This type of dimension is used in the establishment of relations for **case study** sections. You may use these dimensions in addition to the normal dimensions to solve your sections.

Dimensions - Creating

Angular

Used to create dimensions that measure the angle between two sketched lines. Also used to create dimensions that measure the angle between the endpoints of an arc.

To Create an Angular Dimension (between Two Sketched Lines)

❶ Select **Dimensions** from the SKETCHER menu.

❷ Choose the lines to dimension, using the **left** mouse button.

❸ Place the dimension at the desired location, using the **middle** mouse button.

(**Where** you place the dimension determines if the angle measured is **acute** or **obtuse**.

To Create an Angular Dimension (between Arc Endpoints)

❶ Select **Dimensions** from the SKETCHER menu.

❷ Choose one arc endpoint to dimension, using the **left** mouse button.

❸ Choose the other arc endpoint to dimension, using the **left** mouse button.

❹ Choose the arc element along the body of the arc, using the **left** mouse button.

❺ Place the dimension at the desired location, using the **middle** mouse button.

Diameter

Used to create dimensions that measure the diameter of sketched circles or arcs. Also used to create dimensions that measure the diameter of sketched sections, that will be revolved about a centerline.

To Create a Diameter Dimension (for Arcs or Circles)

❶ Select **Dimensions** from the SKETCHER menu.

❷ Choose the arc or circle to dimension, using the **left** mouse button.

❸ *AGAIN* Choose the arc or circle to dimension, using the **left** mouse button.

❹ Place the dimension at the desired location, using the **middle** mouse button.

To Create a Diameter Dimension (for Sketched Elements to be Revolved by a Centerline)

❶ Select **Dimensions** from the SKETCHER menu.

❷ Choose the sketched element to dimension, using the **left** mouse button.

❸ Choose the centerline to be used as the revolution axis, using the **left** mouse button.

❹ *AGAIN* Choose the sketched element to dimension, using the **left** mouse button .

❺ Place the dimension at the desired location, using the **middle** mouse button.

Linear

Used to create dimensions that measure the length of a sketched line segment or the distance between two elements.

To Create a Linear Dimension (for a Sketched Line Segment)

❶ Select **Dimensions** from the SKETCHER menu.

❷ Choose the line to dimension, using the **left** mouse button.

❸ Place the dimension at the desired location, using the **middle** mouse button.

To Create a Linear Dimension (for the Distance between Two Parallel Lines)

❶ Select **Dimensions** from the SKETCHER menu.

❷ Choose both the lines to dimension, using the **left** mouse button.

❸ Place the dimension at the desired location, using the **middle** mouse button.

To Create a Linear Dimension (for the Distance between a Point and a Line Segment)

❶ Select **Dimensions** from the SKETCHER menu.

❷ Choose the line to dimension, using the **left** mouse button.

❸ Choose the point to dimension, using the **left** mouse button.

❹ Place the dimension at the desired location, using the **middle** mouse button.

To Create a Linear Dimension (for the Distance between Two Points)

❶ Select **Dimensions** from the SKETCHER menu.

❷ Choose both the points to dimension, using the **left** mouse button.

❸ Place the dimension at the desired location using the **middle** mouse button.

❹ Select the type of dimension to place: **Horizontal, Vertical,** or **Shortest Distance** (Slanted).

Spline

Used to create dimensions that measure a sketched spline by it's endpoints or interpolation points.

Ordinate

This command works with the **baseline** dimension command to create dimensions that measure the relative distance from a specified baseline.

To Create an Ordinate Dimension

❶ Select **Dimensions** from the SKETCHER menu.

❷ Choose the baseline dimension, using the **left** mouse button.

❸ Choose the element to be dimensioned, using the **left** mouse button.

❹ Place the dimension at the desired location, using the **middle** mouse button.

Radial

Used to create dimensions that measure the radii of sketched circles or arcs. Also used to create dimensions that measure the radii of circles and arcs, that will be revolved about a centerline.

To Create a Radial Dimension (for Arcs or Circles)

❶ Select **Dimensions** from the SKETCHER menu.

❷ Choose the arc or circle to dimension, using the **left** mouse button.

❸ Place the dimension at the desired location, using the **middle** mouse button.

To Create a Radial Dimension (for Sketched Elements to be Revolved about a Centerline)

❶ Select **Dimensions** from the SKETCHER menu.

❷ Choose the sketched element to dimension, using the **left** mouse button.

❸ Choose the centerline to be used as the revolution axis, using the **left** mouse button.

❹ Place the dimension at the desired location, using the **middle** mouse button.

Sketch

SKETCHER ➡ SKETCH

SKETCHER
Restart
Sketch
Pick Curve
Tools
Modify
Regenerate
Delete
Undelete
Dimension
Align
Unalign
Interface
Set Up
Toggle
Relation
Done
Quit

GEOMETRY
Centerline
Tang line
Parl line
Perp line
2 Tang line
Fillet
3 Pt Arc
Concentric
Point
Coord sys
Spline
Conic
Text

2 Tang Lines

Creates a tangent line between two specified arcs, circles, or splines.

3 Point Arc

Creates an arc by defining the beginning and ending points, and then a point that the arc's radius will pass through.

Centerlines

Creates a sketched centerline. These lines are special in that they are used to define either symmetry within a section or the axis of revolution for a revolved feature.

Concentric

Creates arcs or circles that are concentric to existing geometry (part geometry or sketched elements). You create concentric arcs by using this command and the **right** mouse button. To create concentric circles, use this command with the **middle** mouse button.

Conic

Creates a two or three point conic spline based upon the options chosen.

Coord Sys (See Coordinate Systems)

This command is used to create a coordinate system to assist in the creation of sketcher sections.

> *NOTE: In sketcher mode, coordinate systems, like sketched points, are useful in dimensioning elements. They are also useful in establishing a relative origin for each of the sections used to create blended features.*

Fillet

Creates a rounded intersection between two elements. The radius of the fillet element is defined by the selection point closest to the intersection of the two elements. This command automatically clips and deletes unwanted sections of selected lines. When used on other elements, the unwanted sections must be deleted manually.

Parallel Lines

Creates a line parallel to lines, centerlines, datum planes, datum axes, and existing part edges.

Perpendicular Lines

Creates lines perpendicular to lines, centerlines, datum planes, datum axes, and existing part edges.

Point

Places a point at a known location (projected intersection of two filleted elements). These points are very helpful when dimensioning to locations where no geometry exists. They are also helpful in creating splines.

Spline

Creates a spline element. A spline is defined as curves that smoothly pass through a series of given points.

Tangent Lines

Creates a line tangent to an existing arc or spline element.

Text

Used to place text in a section. The text can either be cosmetic or part of an extruded protrusion / cut. For cosmetic text, you may use any font (you must use the **Modify Text** commands after placing the text), for characters in an extruded feature, you must use the default font (FONT3D).

SKETCHER ➡ SKETCH ➡ CONIC

SKETCHER
Restart
Sketch
Pick Curve
Tools
Modify
Regenerate
Delete
Undelete
Dimension
Align
Unalign
Interface
Set Up
Toggle
Relation
Done
Quit

GEOMETRY
Centerline
Tang line
Parl line
Perp line
2 Tang line
Fillet
3 Pt Arc
Concentric
Point
Coord sys
Spline
Conic
Text

CONIC PNTS
Two Points
Three Points
Known Point
Sketch Point
Done Conic
Quit Conic

Known Point

Used in three point conic spline creation. This command is used to select a model vertex or datum point which is on the model. This command is only available in the 3D sketcher.

Sketch Point

Used with three point conic spline creation. This command is used to select a sketched point or entity vertex. The entity vertex may **not** be a conic or spline endpoint. If this is required, use the sketcher **Point** command to create a sketched point at the entity's endpoint, then use the sketched point to create the conic.

Three Points

Creates a three point conic spline using sketched or known points.

Two Points

Creates a two point conic spline from user specified points.

SKETCHER ➥ SKETCH ➥ SPLINE

SKETCHER
Restart
Sketch
Pick Curve
Tools
Modify
Regenerate
Delete
Undelete
Dimension
Align
Unalign
Interface
Set Up
Toggle
Relation
Done
Quit

GEOMETRY
Centerline
Tang line
Parl line
Perp line
2 Tang line
Fillet
3 Pt Arc
Concentric
Point
Coord sys
Spline
Conic
Text

SPLINE MODE
Sketch Pnts
Select Pnts
Done / Return

Select Points

Creates a spline element by selecting existing sketcher points. The points selected are only used to place the spline, and no long term association exists.

Sketch Points

Creates a spline element by arbitrarily placing points for the spline to pass through.

SKETCHER ➥ SKETCH ➥ SPLINE ➥ TANGENCY

SKETCHER
Restart
Sketch
Pick Curve
Tools
Modify
Regenerate
Delete
Undelete
Dimension
Align
Unalign
Interface
Set Up
Toggle
Relation
Done
Quit

GEOMETRY
Centerline
Tang line
Parl line
Perp line
2 Tang line
Fillet
3 Pt Arc
Concentric
Point
Coord sys
Spline
Conic
Text

SPLINE MODE
Sketch Pnts
Select Pnts
Done / Return
TANGENCY
None
Start
End
Both

Both

Used with the spline creation command. Instructs Pro/ENGINEER to create a spline with both endpoints tangent.

End

Used with the spline creation command. Instructs Pro/ENGINEER to create a spline with the second endpoint tangent.

None

Used with the spline creation command. Instructs Pro/ENGINEER to create a spline with neither endpoint tangent.

Start

Used with the spline creation command. Instructs Pro/ENGINEER to create a spline with the first endpoint tangent.

Interface

SKETCHER ➡ INTERFACE

SKETCHER
Restart
Sketch
Pick Curve
Tools
Modify
Regenerate
Delete
Undelete
Dimension
Align
Unalign
Interface
Set Up
Toggle
Relation
Done
Quit

INTERFACE
Import
Export

Export

Provides access to plotting and file export capabilities.

Import

Imports an IGES file containing 2D geometry to be used as a sketched section. In most cases, the geometry will need to have dimensional values added.

Modify

SKETCHER ➥ MODIFY ➥ SPLINE

SKETCHER
Restart
Sketch
Pick Curve
Tools
Modify
Regenerate
Delete
Undelete
Dimension
Align
Unalign
Interface
Set Up
Toggle
Relation
Done
Quit

MOD SPLN
Move pnts
Coords
Read pnts
Save pnts
Info pnts
Add pnts
Delete pnts
Tangency
Done Modify
Quit Modify

NOTE: These options appear only when a spline is selected with the Modify command.

Add Pnts

This command is used to insert a new spline point between two existing spline points.

Coords

This command is only available when the spline element has been associated to a sketcher coordinate system. This command allows you to enter new X and Y coordinate values for an existing point via the keyboard. The type of coordinate system can be modified when you use the **Save** or **Read** points commands, but only Cartesian coordinate values can be entered using this command.

Delete Pnts

This command is used to erase an existing spline point.

Info Pnts

This command is used to display the coordinate values of an existing spline. You will be prompted for the type of coordinate system in which to display the values.

Move Pnts

This command is used to interactively relocate the points of an existing spline element.

Read Pnts

This command is only available when the spline element has been associated to a sketcher coordinate system. This command allows you to enter new X and Y coordinate values for an existing spline from a specified text file.

Save Pnts

This command is only available when the spline element has been associated to a sketcher coordinate system. This command allows you to save X and Y coordinate values for an existing spline into a specified text file.

Tangency

This command is used to modify the tangency conditions of an existing spline element's beginning or ending points.

SKETCHER ➡ MODIFY ➡ SPLINE ➡ TANGENCY

SKETCHER
Restart
Sketch
Pick Curve
Tools
Modify
Regenerate
Delete
Undelete
Dimension
Align
Unalign
Interface
Set Up
Toggle
Relation
Done
Quit

MOD SPLN
Move pnts
Coords
Read pnts
Save pnts
Info pnts
Add pnts
Delete pnts
Tangency
Done Modify
Quit Modify

MOD TANG
Add
Remove

Add

This command is used to create a tangency condition at the beginning or ending point of an existing spline element.

Remove

This command is used to delete a tangency condition at the beginning or ending point of an existing spline element.

SKETCHER ➡ MODIFY ➡ TEXT

SKETCHER
Restart
Sketch
Pick Curve
Tools
Modify
Regenerate
Delete
Undelete
Dimension
Align
Unalign
Interface
Set Up
Toggle
Relation
Done
Quit

MOD SEC TEXT
Text Line
Angle
Text Height
Text Width
Slant Angle
Font
Done Modify
Quit Modify

NOTE: These options appear only when text is selected with the Modify command.

Angle

Used to specify a value for the orientation angle of the selected text. A positive value rotates the text in a counter-clockwise direction. A value of **-1** will return the text to the default orientation.

Font

Used to specify a new font name for the selected text.

- Font3D — The default font used for all extruded text and text features
- Font — ASCII-based font characters
- Leroy — Characters matching the leroy standard
- Cal_alf — Calcomp alpha-numeric characters
- Cal_grek — Calcomp Greek characters

Slant Angle

Used to specify a new value for the slant (if any) of the selected text's characters.

Text Height

Used to specify a new value for the height of the selected text's characters.

Text Line

Used to edit the characters used in a line of text.

Text Width

Used to specify a new value for the width of the selected text's characters.

Setup

SKETCHER ➥ SETUP

SKETCHER
Restart
Sketch
Pick Curve
Tools
Modify
Regenerate
Delete
Undelete
Dimension
Align
Unalign
Interface
Set Up
Toggle
Relation
Done
Quit

SET UP
Num Digits
Accuracy
Declaration

Accuracy

This command works with the sketcher **Regenerate** command to solve sketched sections. Valid accuracy values range from 1.0e-9 (0.000000001) to 1.0 (the default).

Declaration

This command works with **case study** sections used in **Layout** mode. (See Section Four DRAWING.)

Number Digits

This command determines how many decimal values will be displayed for all sketcher dimensions. The default is 2. You may specify anywhere from 0 to 14 decimals be displayed.

Tools

SKETCHER ➡ TOOLS

SKETCHER
Restart
Sketch
Pick Curve
Tools
Modify
Regenerate
Delete
Undelete
Dimension
Align
Unalign
Interface
Set Up
Toggle
Relation
Done
Quit

SEC TOOLS
Intersect
Trim
Start pnt
Move Dim
Use edge
Offset edge
Sec Info
Mirror
Cosm Font
Copy Layout
Replace
Thicken
Grid

Copy Layout

Copies undimensioned geometry from a layout sketch to the **case study** window.

Edge — Offset

Creates sketch geometry by copying edge definitions from the part geometry to the sketching plane. Unlike **Edge Use**, this command can relocate the copied geometry a specified distance from the original part geometry.

Edge — Use

Creates sketch geometry by copying edge definitions from the part geometry to the sketching plane.

Grid

This option provides access to the tools used to modify the sketcher's grid display.

Intersect

Finds the intersection of two selected elements that cross each other, and divides each element at that point. When used to divide a circle (into two separate arcs) where a line crosses, the circle must have the command run at both intersections.

Mirror

Used to create a reciprocal copy of selected sketched elements about a sketched centerline.

Move Dim

Used to move dimension text from one location to another. As the text is relocated, any leader lines or other dimension elements are also adjusted.

Replace

This command is used to replace existing sketched geometry with newly sketched geometry.

Sec Info

Used to display various types of information about a selected element.

Start Point

This command is used to reorient the start point of a sketched section and is most often used to align start points of two sections for creating blend features.

Trim

Used to trim a single element to a point controlled by a bounding entity, a specified length, or an incremental value. The trimming process can shorten or lengthen an element, as is required to satisfy the variables chosen.

NOTE: Pieces of elements deleted by the **Trim** *command cannot be undeleted.*

Tools - Grid

SKETCHER ➥ TOOLS ➥ GRID

SKETCHER
Restart
Sketch
Pick Curve
Tools
Modify
Regenerate
Delete
Undelete
Dimension
Align
Unalign
Interface
Set Up
Toggle
Relation
Done
Quit

SEC TOOLS
Intersect
Trim
Start pnt
Move Dim
Use edge
Offset edge
Sec Info
Mirror
Cosm Font
Copy Layout
Replace
Thicken
Grid

GRID MODIFY
Grid On
Grid Off
Type
Origin
Grid Params

Grid Off

Turns **off** the grid display.

Grid On

Turns **on** the grid display.

Grid Params

Allows you to specify the spacing and angles of various grid variables for both Cartesian and Polar grids.

Origin

Allows you to relocate the grid's intersection point. The acceptable locations are:

- Sketched points
- Datum points
- Vertices of curves or edges
- The endpoint or midpoint of a sketched element

Type

Allows you to specify a grid type of **Cartesian** or **Polar**.

SKETCHER ➡ TOOLS ➡ GRID ➡ GRID PARAMS (Cartesian Options)

SKETCHER
Restart
Sketch
Pick Curve
Tools
Modify
Regenerate
Delete
Undelete
Dimension
Align
Unalign
Interface
Set Up
Toggle
Relation
Done
Quit

SEC TOOLS
Intersect
Trim
Start pnt
Move Dim
Use edge
Offset edge
Sec Info
Mirror
Cosm Font
Copy Layout
Replace
Thicken
Grid

GRID MODIFY
Grid On
Grid Off
Type
Origin
Grid Params

CART PARAMS
X & Y Spacing
X Spacing
Y Spacing
Angle

Angle

Specifies the angle between horizontal and the X grid axis (a rotated grid pattern).

X Spacing

Sets the spacing for the X grid value only.

X and Y Spacing

Simultaneously sets the X and Y grid spacing to a specified value.

Y Spacing

Sets the spacing for the Y grid value only.

SKETCHER ➥ TOOLS ➥ GRID ➥ GRID PARAMS (Polar Options)

SKETCHER
Restart
Sketch
Pick Curve
Tools
Modify
Regenerate
Delete
Undelete
Dimension
Align
Unalign
Interface
Set Up
Toggle
Relation
Done
Quit

SEC TOOLS
Intersect
Trim
Start pnt
Move Dim
Use edge
Offset edge
Sec Info
Mirror
Cosm Font
Copy Layout
Replace
Thicken
Grid

GRID MODIFY
Grid On
Grid Off
Type
Origin
Grid Params

POLAR PARAMS
Ang Spacing
Num Lines
Rad Spacing
Angle

Angle

Specifies the angle between horizontal and the 0-degree grid axis (a rotated grid pattern).

Ang Spacing

Specifies the angular spacing between radial lines. The value must divide evenly into 360.

Num Lines

Specifies the quantity of radial lines in the grid pattern, based upon the formula 360/number of lines.

Rad Spacing

Specifies the spacing of the circular grid indicators.

SKETCHER ➡ TOOLS ➡ GRID ➡ TYPE

SKETCHER
Restart
Sketch
Pick Curve
Tools
Modify
Regenerate
Delete
Undelete
Dimension
Align
Unalign
Interface
Set Up
Toggle
Relation
Done
Quit

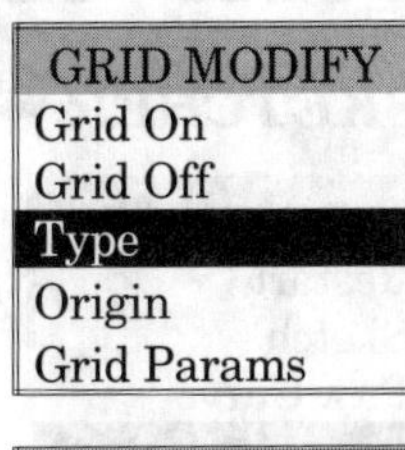

GRID TYPE
Cartesian
Polar

Cartesian

Instructs Pro/ENGINEER to display a **Cartesian** type of rectangular grid.

Polar

Instructs Pro/ENGINEER to display a **Polar** type of radial grid.

Tools - Mirror

SKETCHER ➥ TOOLS ➥ MIRROR

SKETCHER
Restart
Sketch
Pick Curve
Tools
Modify
Regenerate
Delete
Undelete
Dimension
Align
Unalign
Interface
Set Up
Toggle
Relation
Done
Quit

SEC TOOLS
Intersect
Trim
Start pnt
Move Dim
Use edge
Offset edge
Sec Info
Mirror
Cosm Font
Copy Layout
Replace
Thicken
Grid

MIRROR
Pick
All
Done
Quit

All

Used to simultaneously select all sketched elements, to be mirrored.

Pick

Used to select individual elements, to be mirrored.

Tools - Section Info

SKETCHER ➥ TOOLS ➥ SEC INFO

SKETCHER
Restart
Sketch
Pick Curve
Tools
Modify
Regenerate
Delete
Undelete
Dimension
Align
Unalign
Interface
Set Up
Toggle
Relation
Done
Quit

SEC TOOLS
Intersect
Trim
Start pnt
Move Dim
Use edge
Offset edge
Sec Info
Mirror
Cosm Font
Copy Layout
Replace
Thicken
Grid

SEC INFO
Entity
Intrsct pt
Tangent pt
References
No Csys
X-Angle
Angle
Distance
Grid Info
Done / Return

Angle

Used to measure the angle between any two selected line segments.

Distance

Used to measure the distance between a point and a line segment, two points, or two parallel lines.

Entity

This command displays information about a selected element. The information includes type of geometry and endpoint tangencies. Endpoint coordinates will also be displayed if a coordinate system is chosen.

Grid Info

This command displays the current grid spacing and angle values in the message window.

Intersect Pt

This command determines the intersection point(s) of two selected elements. The elements must intersect or no point will be found.

No Csys

This command instructs Pro/ENGINEER **not** to calculate any coordinate points. You will be prompted for this command at the appropriate time.

References

This command highlights the existing part geometry or the datums that were used as references to locate the current sketched section. The colors used are **blue** for edges, **yellow** for surfaces and **red** for datums and features.

Tangent Pt

This command displays the distance between the tangent points of two selected elements (they do not have be touching). It also displays the angle of slope at the tangency points, and, if a coordinate system is selected, the coordinates of the tangent points.

Tools - Trim

SKETCHER ➥ TOOLS ➥ TRIM — Options

SKETCHER
Restart
Sketch
Pick Curve
Tools
Modify
Regenerate
Delete
Undelete
Dimension
Align
Unalign
Interface
Set Up
Toggle
Relation
Done
Quit

SEC TOOLS
Intersect
Trim
Start pnt
Move Dim
Use edge
Offset edge
Sec Info
Mirror
Cosm Font
Copy Layout
Replace
Thicken
Grid

OPTIONS
Bound
Corner
Length
Increm

Bound

Trims an existing element to a point controlled by another element. The element selection point determines which end of the element will be modified accordingly. If the element is to be reduced, choose the portion of the element to keep.

Increm

Trims an existing element by a specified value. A positive value lengthens the element, while a negative value reduces the element's length. The element selection point determines which end of the element will be modified accordingly.

Length

Trims an existing element to a specified value. If the value is greater than the element's current length, the element will be extended. If the value is less than the element's current length, the element will be reduced. The element selection point determines which end of the element will be modified accordingly.

Section Three

DATUMS

Datum Commands

MODE ➡ PART ➡ FEATURE ➡ CREATE ➡DATUM

MODE
Sketcher
Part
Sheet Metal
Composite
Assembly
Drawing
Manufacture
Mold
Layout
Format
Report
Markup
Diagram

PART
Feature
Modify
Regenerate
Relations
Family Tab
Interchange
Declare
Info
Interface
Set Up
Ref Dim
X-Section
Layer
Program

FEAT
Create
Pattern
Copy
Delete
Del Pattern
Group
Supress
Resume
Reorder
Read Only
Redefine
Reroute
Mirror Geom
Insert Mode
Done

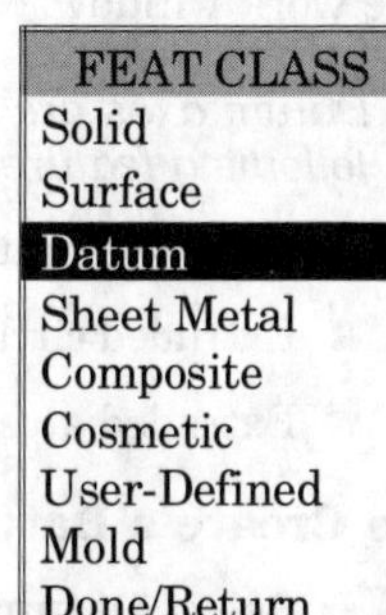

DATUM
Plane
Axis
Curve
Point
Coord Sys
Graph

Normally, you can place only one datum feature (planes, axes, points, curves) at a time, then you are required to repeat all the menu picks to place another. There is an option you can set in the CONFIG.PRO file to allow repeated placement of datum features. Set repeat_datum_create to yes using the MISC ➡ EDIT CONFIG.PRO menu option.

Creating

Axis

Used as a reference for feature creation. Most commonly used for placement of datum planes, placing features concentrically, and creating radial patterns of features. Datum axes appear as centerlines in the work window.

Datum axes are automatically created by Pro/ENGINEER for the following features:

- Revolved features
- Extruded circles
- Extruded arcs

To Create a Datum Axis

❶ Select **Datum** from the FEATURES menu, and then select **Axis** .

❷ Choose the type of axis placement you will be using (single or patterned), then select **Done** .

❸ Choose the type of *placement constraint* you will be using to locate the datum axis.

❹ Choose the necessary references.

Curve

A datum curve is used to create datum surfaces, and as a trajectory for sweeps. Datum curves are also used in manufacturing.

To Create a Datum Curve

❶ Select **Datum** from the FEATURES menu, then select **Curve**.

❷ Choose the type of curve placement you will be using (sketched, single, or patterned) then select **Done.**

❸ Choose the type of *placement constraint* you will be using to locate the datum curve.

❹ Choose the necessary references.

Coord Sys

Coordinate systems are established arrangements of directional values, used to reference features. In Pro/ENGINEER they are used in mass property calculations, component placement in assemblies, meshing, and manufacturing.

Coordinate systems are always displayed with X,Y and Z axes. However, when using the coordinate system as a reference to create new features, the information can be interpreted according to any of three standard systems:

- Cartesian: X-value, Y-value, Z-value
- Cylindrical: Radius, Theta, Z-value
- Spherical: Radius, Theta, Phi

Graph

A graph is a special type of sketched feature, used to associate a function with a part. Graphs are used most often in multi-trajectory sweep relations. A graph feature is not displayed in the model, it is only accessible in the **Part Info**.

Planes

Datum planes are features used to create a reference on a part where one does not already exist.

When appropriate, Pro/ENGINEER allows you to create datum planes using the Make Datum option. Datum planes that are created using this option become embedded in the feature created. They are invisible after the feature is completed. Dimensional constraints used to reference the embedded datum plane will appear, when modifying the feature, as part of the feature's dimensional values.

NOTE: Datum planes have two sides: yellow and red. These colors are designed to assist you when assembling parts or orienting views. The color of the plane in the work window will depend upon which side of the datum plane is currently facing the screen.

To Create a Datum Plane

Select **Datum** from the FEAT CLASS menu, then select **Plane** (or **Make Datum** from the SETUP PLANE sub-menu).

❶ Choose the type of placement you will be using (single or patterned), select the appropriate *sizing option*, then select **Done.**

❷ Choose the type of *placement constraint* you will be using to locate the datum plane.

❸ Choose the necessary references.

❹ Repeat steps two and three until all the desired constraints have been established.

❺ Choose **Done** to create the datum plane. (See Datum Plane — options.)

NOTE: If Pro/ENGINEER cannot create the datum plane as specified (due to conflicting or insufficient constraints) it will automatically activate the **Redefine** *command. Choose* **Attributes** *and then* **Done** *to make another attempt at creating the datum plane.*

Point

Datum points are used to clearly designate points in space. These known points can be used to place datum planes, axes, trajectories, holes and shafts. Datum points are also used as load points in mesh generation.

Datum points appear as the character "X" followed by the abbreviation **PNTn**, where **n** is the datum point number.

To Create a Datum Point

❶ Select **Datum** from the FEAT CLASS menu, then select **Point**.

❷ Choose the type of placement you will be using (single or patterned).

❸ Choose the type of *placement constraint* you will be using to locate the datum point then select **Done**.

❹ Choose the necessary references.

Axis

DATUMS ➡ AXIS

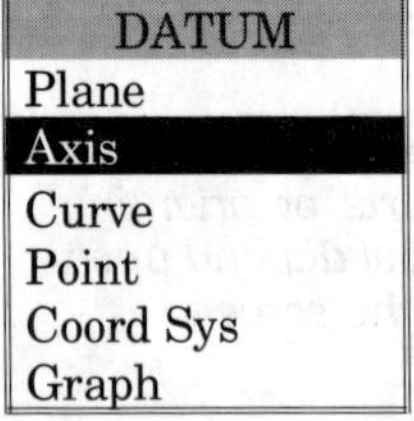

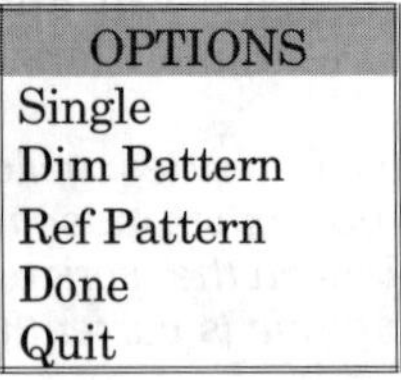

Dim Pattern

Create a pattern of datum axes using specified dimensional values to control the pattern.

Ref Pattern

Create a pattern of datum axes that are dependent upon an existing **Dim Pattern** for their dimensional values.

NOTE: When Pro/ENGINEER determines it is not logical to have both a Dim Pattern and Ref Pattern, only the option **Pattern** *will* *be visible. In these cases,* **Pattern** *has the same function as* **Dim Pattern**.

Single

Creates a single datum axis. This is the default menu option.

DATUMS ➡ AXIS ➡ DONE

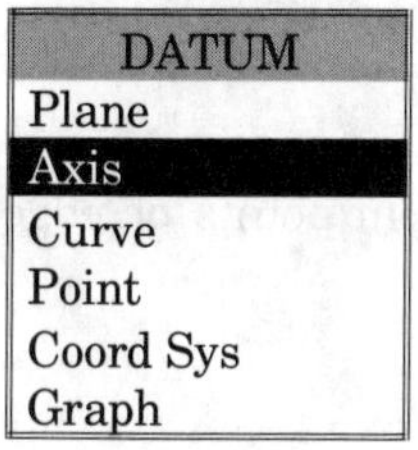

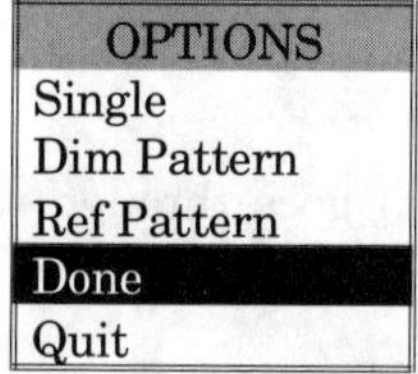

DATUM AXIS
Thru Edge
Normal Pln
Pnt Norm Pln
Thru Cyl
Two Planes
Two Pnt / Vtx
Pnt on Surf
Pnt on Surf
Tan Curve
Done / Quit

Normal Pln

Creates a datum axis that is normal to a selected surface, and includes linear dimensions from the edges of the surface to locate the axis.

Pnt — Norm Pln

Creates a datum axis that goes through a datum point, and is normal to a selected plane.

Pnt — On Surf

Creates a datum axis that goes through an *on surface* datum point, and is normal to the surface at that point.

Tan Curve

Creates a datum axis that is tangent to the endpoint of an edge or curve. Select the feature, then select which end of the feature to be tangent to.

Thru Cyl

Creates a datum axis through the *assumed* axis of any surface of revolution. Select a cylindrical or revolved surface that does not have an existing axis.

Thru Edge

Creates a datum axis along an edge of your part. Select the edge to use.

Two Planes

Creates a datum axis at the *intersection* of two planes (datum planes or surfaces). Select two planes that either *do* intersect each other or eventually *would* (projected intersection).

Two Pnt / Vtx

Creates a datum axis that goes through two datum points or edge vertices. Select the desired points or vertices.

Coordinate System

DATUMS ➡ COORD SYS

DATUM
Plane
Axis
Curve
Point
Coord Sys
Graph

OPTIONS
3 Planes
Pnt + 2 Axes
2 Axes
Offset
Offs By View
Pln + 2 Axes
Orig + ZAxis
From File
Default
Done
Quit

2 Axes

Creates a coordinate system, with the origin being at the intersection of two selected axes, and defines the orientation of a plane through the origin and the axis selected first.

3 Planes

Creates a coordinate system, with the origin being at the intersection of the three planes selected. The planes can be datum planes or planar surfaces. The planes need not be orthogonal. The first coordinate system axis is defined by the first selected plane's normal. The second axis is defined by the normal of the plane selected second, and the third axis is defined by the right hand rule.

Default

Creates a default coordinate system for the part. The origin of the coordinate system will be at the first sketched vertex of the base feature. The X-axis will be defined by the sketched horizontal, the Y-axis by the sketched vertical, and the Z-axis by the right hand rule.

From File

Creates a coordinate system using a data file to define a transformation matrix. The matrix locates the new coordinate system relative to an existing coordinate system.

Offset

Creates a coordinate system that is offset from an existing coordinate system.

Plane + 2 Axes

Creates a coordinate system, with the origin being at the intersection of the selected plane and the first selected axis. The third point defines the orientation of a plane through the origin and the axis selected first.

Point + 2 Axes

Creates a coordinate system, with the origin being at the first selected point. The second point defines a direction for one of the coordinate system axes. The third point defines the orientation of a plane through the origin and the axis selected first.

DATUMS ➥ COORD SYS ➥ 3 PLANES or PNT + 2 AXES or PLN + 2 AXES

DATUM
Plane
Axis
Curve
Point
Coord Sys
Graph

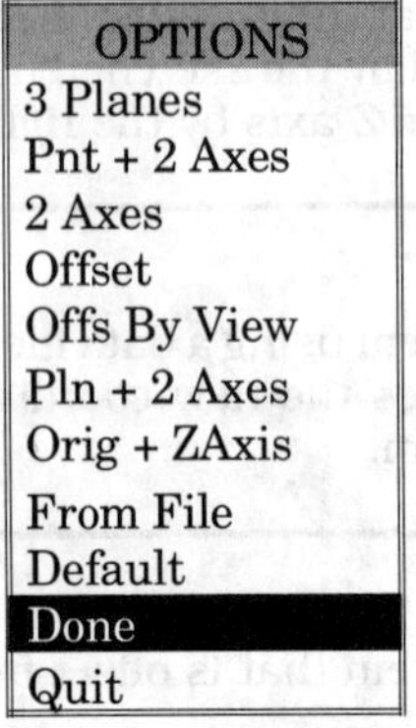

SET AXIS
Entity / Edge
Plane Norm
2 Points
Orig + Pnt
Quit

COORD SYS
X-Axis
Y-Axis
Z-Axis
Next
Previous
Reverse

2 Points

Used to select two points which define a vector.

Entity / Edge

Used to select a datum axis, straight edge, or straight curve.

Orig + Point

Works with the coordinate system origin and a selected point to define a vector.

Plane Normal

Used to select the normal of an identified plane.

Reverse

Used during coordinate system creation to flip the red directional arrow prior to axis definition.

X — axis

Used during coordinate system creation to define which axis will serve as the X-axis.

Y — axis

Used during coordinate system creation to define which axis will serve as the Y-axis.

Z — axis

Used during coordinate system creation to define which axis will serve as the Z-axis.

DATUMS ➥ COORD SYS ➥ OFFSET

DATUM
Plane
Axis
Curve
Point
Coord Sys
Graph

OPTIONS
3 Planes
Pnt + 2 Axes
2 Axes
Offset
Offs By View
Pln + 2 Axes
Orig + ZAxis
From File
Default
Done
Quit

MOVE
Translate
Rotate
Done Move
Quit Move

Rotate

Specifies a rotation angle from an existing coordinate system.

Translate

Specifies an offset value from an existing coordinate system .

DATUMS ➥ COORD SYS ➥ OFFSET ➥ TRANSLATE or ROTATE

DATUM
Plane
Axis
Curve
Point
Coord Sys
Graph

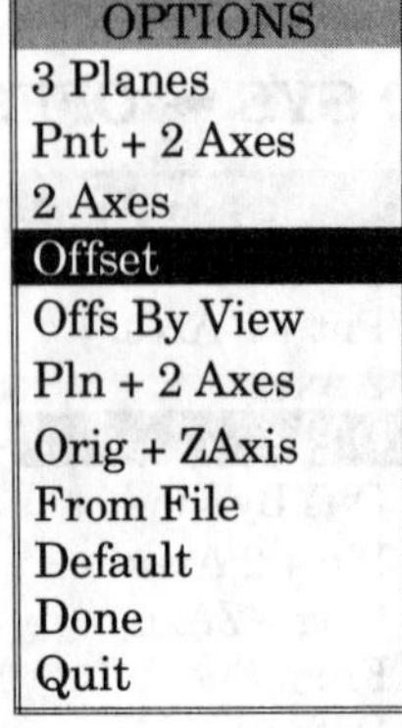

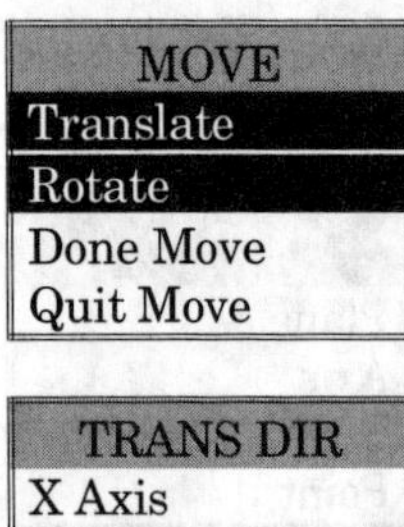

TRANS DIR
X Axis
Y Axis
Z Axis

ROTATE DIR
X Axis
Y Axis
Z Axis

X — axis

Used during coordinate system creation to define which axis will serve as the X-axis.

Y — axis

Used during coordinate system creation to define which axis will serve as the Y-axis.

Z — axis

Used during coordinate system creation to define which axis will serve as the Z-axis.

Curves

DATUMS ➥ CURVE

DATUM
Plane
Axis
Curve
Point
Coord Sys
Graph

OPTIONS
Single
Dim Pattern
Ref Pattern
Xhatch
No Xhatch
Sketch
Intr. Surfs
Thru Points
From File
From Edges
Composite
Use Xsec
Projected
Formed
Split
Offset
Done
Quit

Composite

Creates a datum curve by combining several datum curves or part edges. This option places you in a special sketcher session, where you select datums curves and/or edges.

NOTE: Composite curves are not automatically assigned names. Use the SET-UP ➥ NAME ➥ FEATURE menu options to name a composite datum curve.

Dim Pattern

Creates a pattern of datum curves, using specified dimensional values to control the pattern.

Formed

Sketches a datum curve that will be projected onto an existing part surface or surface feature, with the length of the original curve segments preserved (no distortion allowed).

From Edges

Creates a datum curve from adjacent part edges or surface feature edges. The selected edges must form a closed loop or a continuous chain.

From File

Imports a datum curve from another file (Pro/ENGINEER's .IBL format). Datum curves created with this option may be redefined, trimmed, or split with other imported curves.

Intr. Surfs

Creates a datum curve at the intersection of any surface features, part surfaces, or datum planes.

Projected

Sketches or selects a datum curve that will be projected onto an existing part surface or surface feature (distortion allowed).

Ref Pattern

Creates a pattern of datum axes that are dependent upon an existing **Dim Pattern** for their dimensional values.

Single

Creates a single datum curve. This is the default menu option.

Sketch

Creates a datum curve from scratch, using the standard sketching utilities.

Thru Points

Creates a datum curve that goes through selected datum points or a datum point array. The resulting datum curve can be created as a single spline, or as a composite of arcs and line segments.

Use Xsec

Creates a datum curve at the intersection of a user-specified cross section and a part outline.

DATUMS ➥ CURVE ➥ FORMED

DATUM
Plane
Axis
Curve
Point
Coord Sys
Graph

OPTIONS
Single
Dim Pattern
Ref Pattern
Xhatch
No Xhatch
Sketch
Intr. Surfs
Thru Points
From File
From Edges
Composite
Use Xsec
Projected
Formed
Split
Offset
Done
Quit

SURFACE LIST
Solid Surfs
Quilt Surfs
Quit

Solid Surfs

Sketches a datum curve on a plane, then projects the curve onto a solid part surface.

Quilt Surfs

Sketches a datum curve on a plane, then projects the curve onto a surface feature.

DATUMS ➡ CURVE ➡ FROM EDGES

DATUM
Plane
Axis
Curve
Point
Coord Sys
Graph

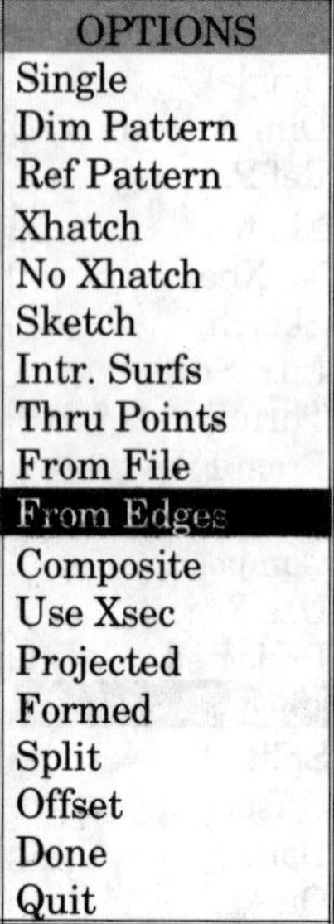

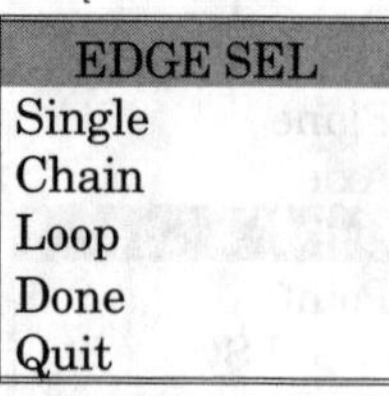

Single

Selects adjacent edges (in any order).

Chain

Selects a single chain of tangent edges.

Loop

Selects a surface with a contained loop. If more than one loop exists, you may use the CONFIRM menu options to make selective choices.

DATUMS ➡ CURVE ➡ INTR. SURFS

DATUM
- Plane
- Axis
- **Curve**
- Point
- Coord Sys
- Graph

INTR SURFS
- Single
- Whole
- Done
- Quit

Single

Selects single surfaces individually. Repeat this option until all the desired surfaces have been selected.

Whole

Selects a whole surface feature or all the surfaces of an existing part.

DATUMS ➡ CURVE ➡ PROJECTED

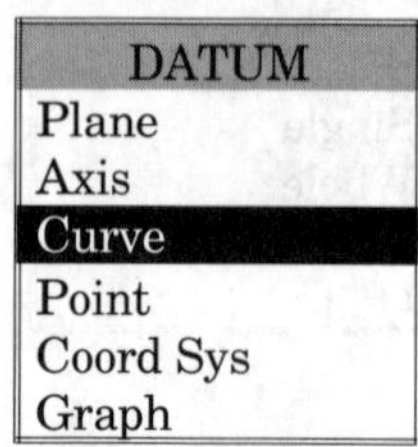

SEC3D OPT
Sket On Pln
Pick 3d Crv
Quit

Sket On Pln

Sketches a datum curve on a plane, then projects the curve onto a surface. The surface must be in a direction which is normal to the sketching plane.

Pick 3D Crv

Selects existing datum curve segments or part edges and projects the resulting datum curve onto a surface. The surface must be in a direction which is normal to a reference plane or tangent to a selected edge, axis, or curve.

DATUMS ➥ CURVE ➥ SKETCH

DATUM
Plane
Axis
Curve
Point
Coord Sys
Graph

OPTIONS
Single
Dim Pattern
Ref Pattern
Xhatch
No Xhatch
Sketch
Intr. Surfs
Thru Points
From File
From Edges
Composite
Use Xsec
Projected
Formed
Split
Offset
Done
Quit

X-Hatch

Cross-hatches the area bounded by a closed loop.

X-Hatch NO

Do not create a cross-hatch pattern.

DATUMS ➡ CURVE ➡ THRU POINTS

DATUM
Plane
Axis
Curve
Point
Coord Sys
Graph

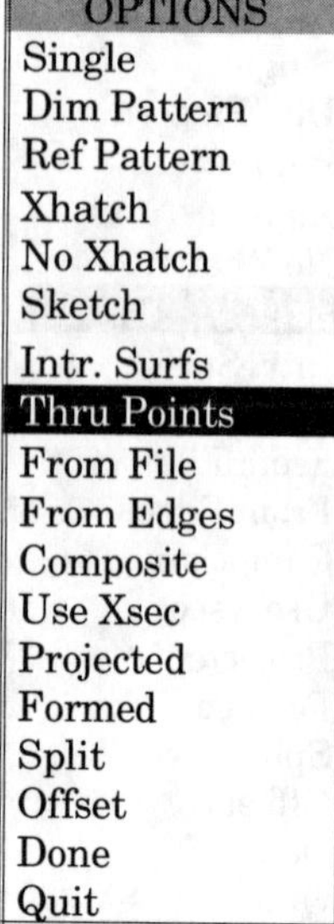

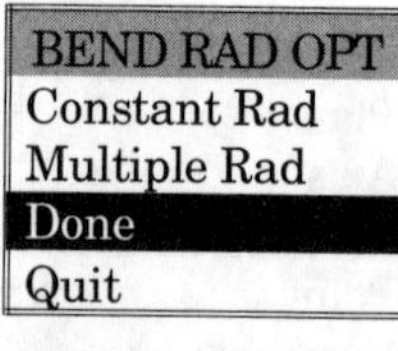

CONNECT TYPE
Line/Arc
Spline
Single Point
Whole Array
Done
Quit

Constant Radius

The bend radius for each of the curve's arc segments will be the same.

Line / Arc

Creates a datum curve consisting of several line and arc segments, beginning and ending with straight lines.

Multiple Radius

The bend radius for each of the curve's arc segments will be individually specified and modifiable.

Single Point

Selects datum points individually. The points can be independent or part of a datum point array.

Spline

Creates a datum curve consisting of a single 3D spline element.

Whole Array

Selects all the datum points in a datum point array in consecutive order.

DATUMS ➥ CURVE ➥ THRU POINTS ➥ SPLINES

DATUM
- Plane
- Axis
- **Curve**
- Point
- Coord Sys
- Graph

OPTIONS
- Single
- Dim Pattern
- Ref Pattern
- Xhatch
- No Xhatch
- Sketch
- Intr. Surfs
- **Thru Points**
- From File
- From Edges
- Composite
- Use Xsec
- Projected
- Formed
- Split
- Offset
- Done
- Quit

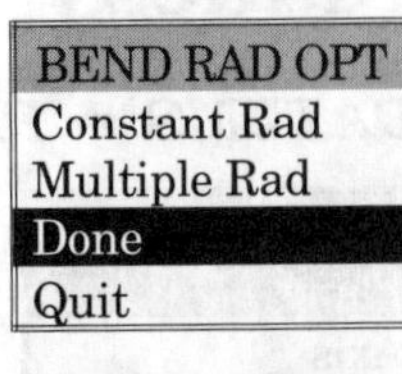

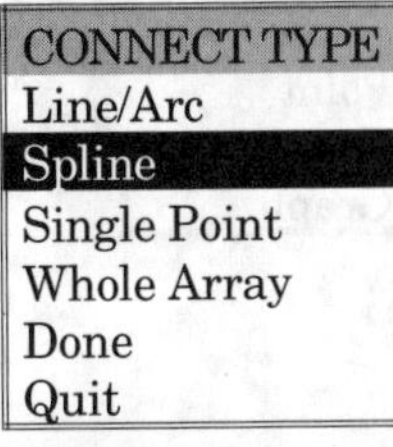

TANGENCY
- None
- Start
- End
- Both

NOTE: These options only become available if the start or end curves are splines.

Both

Both ends of the curve will be tangent to a specified edge or axis.

End

The end of the curve will be tangent to a specified edge or axis.

None

Neither end of the curve will be tangent to a specified edge or axis.

Start

The start of the curve will be tangent to a specified edge or axis.

Planes

DATUMS ➥ PLANE

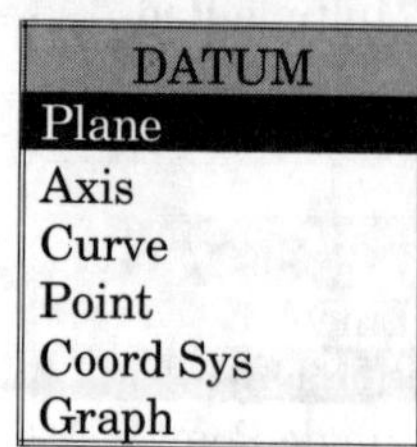

OPTIONS
Default
Fit Part
Fit Feature
Fit Surface
Fit Edge
Fit Axis
Fit Radius
Single
Dim Pattern
Ref Pattern
Done
Quit

NOTE: Datum planes are actually infinite in size. These options control the displayed representation of the datum plane in Pro/ENGINEER's work window. The size of the displayed representation must be associated to an existing feature.

Default

The datum plane will be sized to fit the complete model (part or assembly).

Dim Pattern

Creates a pattern of datum planes using specified dimensional values to control the pattern.

Fit — Axis

The datum plane will be sized to fit an existing axis (feature or datum axis).

Fit — Edge

The datum plane will be sized to fit the edge of a selected feature.

Fit — Feature

The datum plane will be sized to fit a specified feature of a part or assembly.

Fit — Part

The datum plane will be sized to fit a specified part within an assembly. (available in Assembly mode only).

Fit — Radius

The datum plane will be sized to fit a specified radius, centering itself within the constraints of the model.

Fit — Surface

The datum plane will be sized to fit the boundaries of a specified surface.

Ref Pattern

Creates a pattern of datum planes which are dependent upon an existing **Dim Pattern** for their dimensional values.

> *NOTE: When Pro/ENGINEER determines it is not logical to have both a Dim Pattern and Ref Pattern, only the option* **Pattern** *will be visible. In these cases,* **Pattern** *has the same function as* **Dim Pattern**.

Single

Creates a single datum plane. This is the default menu option.

DATUMS ➡ PLANE ➡ DONE

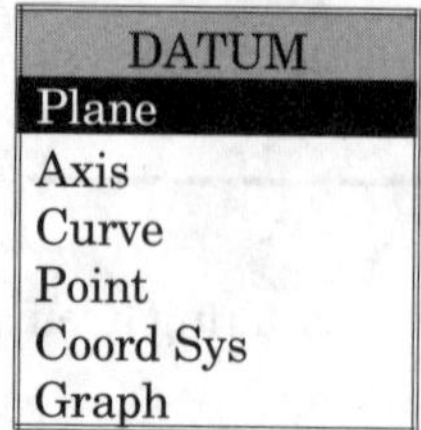

OPTIONS
Default
Fit Part
Fit Feature
Fit Surface
Fit Edge
Fit Axis
Fit Radius
Single
Dim Pattern
Ref Pattern
Done
Quit

DATUM PLANE
Through
Normal
Parallel
Offset
Angle
Tangent
BlendSection
AxisEdgeCurv
Point / Vertex
Plane
Cylinder
Coord Sys
Done
Quit

Angle

Used in combination with other placement constraints (**Through/AxisEdge/Curve**). Specifies an offset angle to control the placement of new datum planes. This command is especially useful when creating patterns of datum planes.

AxisEdgeCurve

Creates a datum plane along the *assumed* axis of a revolved feature or through any straight edge, axis, or curve. If used alone (no prior placement constraint), this option completely constrains the datum plane placement, and the orientation of the new datum plane is determined by Pro/ENGINEER.

> *NOTE: The* **Through/AxisEdgeCurve** *command combination can also be used to create datum planes through imported wireframe geometry and datum curves.*

BlendSection

Creates a datum plane that goes through the section used to create a *blended* feature. If multiple sections exist, you will be prompted for the section number. This option completely constrains the datum plane placement.

Offset/Coord Sys

Creates a datum plane that is normal to one of the coordinate system axes, and is offset from the origin of the coordinate system by a specific distance. This option completely constrains the datum plane placement.

Offset/Plane

Creates a datum plane that is parallel to a specified plane, and is offset by a specific distance. This option completely constrains the datum plane placement.

Through/Cylinder

Creates a datum plane along the *assumed* axis of a revolved feature. If used alone (no prior placement constraint), this option completely constrains the datum plane placement and the orientation of the new datum plane is determined by Pro/ENGINEER.

Through/Plane

Creates a datum plane that is coincident with a specified planar surface. This option completely constrains the datum plane placement.

Point

DATUMS ➥ POINT

DATUM
Plane
Axis
Curve
Point
Coord Sys
Graph

OPTIONS
On Surface
Curve X Srf
On Vertex
Offset Csys
Three Srf
At Center
On Curve
Single
Dim Pattern
Ref Pattern
Done
Quit

At Center

Creates a datum point at the center of an arc or a circle.

Curve X Srf

Creates a datum point at the *intersection* of a curve and a surface.

Dim Pattern

Creates a pattern of datum points using specified dimensional values to control the pattern.

Offset Csys

Creates a datum point array (s single feature consisting of one or more datum points).

On Curve

Creates a datum point on a curve or edge. The point must be located by a dimension along the length of the curve from one of its vertices.

On Surface

Creates a datum point on a specified surface. The datum point must be located with dimensions from two planes or edges.

On Vertex

Creates a datum point on the vertex of a datum curve, imported wireframe, surface feature edge, or part edge.

Ref Pattern

Creates a pattern of datum points that are dependent upon an existing **Dim Pattern** for their dimensional values.

> *NOTE: When Pro/ENGINEER determines it is not logical to have both a Dim Pattern and Ref Pattern, only the option* **Pattern** *will be visible. In these cases,* **Pattern** *has the same function as* **Dim Pattern**.

Single

Creates a single datum point. This is the default menu option.

Three Srf

Creates a datum point at the *intersection* of three surfaces. The surface types can be part surfaces, surface features, or datum planes.

DATUMS ➡ POINT ➡ ON CURVE

DATUM
Plane
Axis
Curve
Point
Coord Sys
Graph

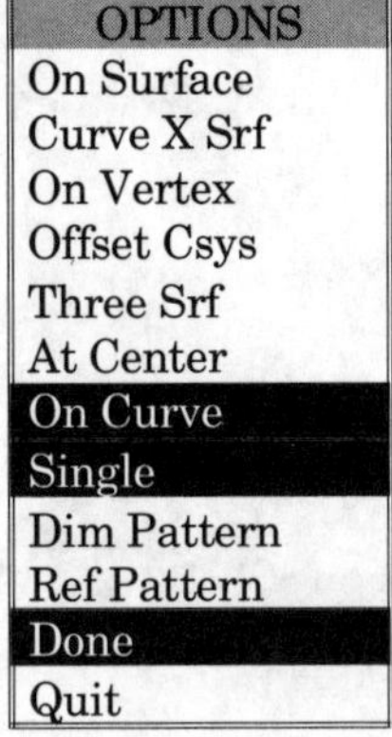

Offset
Lenth Ratio
Actual Length

Offset

Creates a datum point along a curve or an edge, using a planar surface as the reference dimension. This is an exact dimension value.

Length Ratio

Creates a datum point from the curve vertex, using a ratio of the total length as the reference dimension. This method does not use an exact dimension value, but rather a tolerance, as determined by the part accuracy setting.

Actual Len

Creates a datum point from the curve vertex, using a specified distance as the reference dimension. This method does not use an exact dimension value, but rather a tolerance, as determined by the part accuracy setting.

PART

Part Commands

MODE ➥ PART

MAIN	MODE	PART
Mode	Sketcher	Feature
Project	Part	Modify
Dbms	Sheet Metal	Regenerate
Environment	Composite	Relations
Misc	Assembly	Family Tab
Exit	Drawing	Interchange
Quit Window	Manufacture	Declare
ChangeWindow	Mold	Info
View	Layout	Interface
	Format	Set Up
	Report	Ref Dim
	Markup	X-Section
	Diagram	Layer
		Program

Family Table

This command provides access to a very powerful feature of Pro/ENGINEER called **Family Table** or "family of parts." Using this capability you may create a *generic* part that will be used as a template for creating similar (not identical) parts. The dimensions or features of the resulting parts can be different from the original, resulting in, for example, a series of bolts that are the same diameter but get successively longer as you go through the set. The name **Family Table** comes from the fact that the resulting parts are *driven by* a table of dimensional values.

Feature

This command gives you access to all the feature creation utilities.

Layer

Unlike other systems, which use layers in an overlay type of system to separate elements, **layers** are used in Pro/ENGINEER to group various elements together. Once the desired elements have been added to the

layer, you may quickly display or turn off the display of all those grouped elements. You may also suppress features by layer.

Modify

This command displays a sub-menu of various feature modification tools.

Ref Dim

Reference dimensions are dimensions which do not have an impact on part data (they are not driving dimensions). They are placed on a part or drawing for informational purposes only. They are *driven* by the part, and will be automatically adjusted during regeneration.

Regenerate

To regenerate a part is to have Pro/ENGINEER evaluate it's controlling dimensional scheme and solve it. If a part cannot be regenerated, hints will be displayed, prompting you for more information.

Relations

Relations are user defined equations controlling the effect of part modifications. One feature of a part may be related to another, using this capability.

X-section

This command allows you access to all the cross-section creation utilities.

Family Tables

PART ➥ FAMILY TAB

PART	FAMILY TAB
Feature	Add Item
Modify	Delete Item
Regenerate	Set Generic
Relations	Group Table
Family Tab	Edit
Interchange	Show
Declare	Set Type
Info	Instant
Interface	Lock / Unlock
Set Up	Patternize
Ref Dim	Erase Table
X-Section	Show Feat
Layer	Verify
Program	Done / Return

Steps to creating a family table.

❶ Create a *generic* model to be used as the template.

❷ Select the items from the model that will be table driven.

❸ Create the family members by adding instance names and values.

Add Item

This command is used to add items from a generic model to the family table. Use this command if you are creating a new family table or editing an existing table.

Delete Item

This command is used to remove items from a family table.

Edit

This command is used to modify the various components of an existing family table.

Erase Table

This command is used to remove an entire family table from the disk (no undelete).

Group Table

This command is used to create a family table from a user defined feature (UDF). This group may be used as a component in another family table.

Lock / Unlock

This command is a *toggle* switch that changes the current state of a selected instance from **locked** to **unlocked** and vice-versa.

Patternize

This command quickly adds a *range* of instances. For example, you could add all instances in a table ranging from 0.25 to 0.75, with a specified incremental value of 0.125.

Set Generic

This command establishes the selected instance as the *generic* instance. Parts created using the family table that do not have their dimension values modified will be identical to the *generic* instance.

Set Type

This command displays a sub-menu where you choose **Standard** or **Custom** as the family table *type*.

Show

This command instructs Pro/ENGINEER to display the selected family table in a Pro/TABLE window. Using this command, the family table cannot be modified.

Show Feat

This command is used to highlight a feature (in the model) by selecting it from the family table.

PART ➡ FAMILY TAB ➡ ADD ITEM or DELETE ITEM

PART
Feature
Modify
Regenerate
Relations
Family Tab
Interchange
Declare
Info
Interface
Set Up
Ref Dim
X-Section
Layer
Program

FAMILY TAB
Add Item
Delete Item
Set Generic
Group Table
Edit
Show
Set Type
Instant
Lock / Unlock
Patternize
Erase Table
Show Feat
Verify
Done / Return

ITEM TYPE
Dimension
Parameter
Feature
Component
Group
From Menu
Type

NOTE: Selecting either Add Item or Delete Item opens the Item Type menu.

Component

Add or remove an assembly component from the selected family table. This option is only available in **Assembly** mode. The component can be a part or sub-assembly feature.

Dimension

Adds or removes a dimension from the selected family table.

Feature

Adds or removes a feature from the selected family table.

Group

Adds or removes a user defined feature from the selected family table. You may use a UDF in a family table only if the UDF is also *table driven.*

Parameter

Adds or removes a user-defined parameter from the selected family table. Parameters are variables (information) about the part that can be stored without declaring a relationship.

PART ➥ FAMILY TAB ➥ SET TYPE

PART
Feature
Modify
Regenerate
Relations
Family Tab
Interchange
Declare
Info
Interface
Set Up
Ref Dim
X-Section
Layer
Program

FAMILY TAB
Add Item
Delete Item
Set Generic
Group Table
Edit
Show
Set Type
Instant
Lock / Unlock
Patternize
Erase Table
Show Feat
Verify
Done / Return

FAMILY TYPE
Custom
Standard
Quit

Custom

Custom family tables will save *every* instance to disk when you save the generic part.

Standard

Standard family tables save the variables controlling new parts, but do not save *any* instances of the table to disk.

Feature

PART ➥ FEATURE

PART
Feature
Modify
Regenerate
Relations
Family Tab
Interchange
Declare
Info
Interface
Set Up
Ref Dim
X-Section
Layer
Program

FEAT
Create
Pattern
Copy
Delete
Del Patern
Group
Supress
Resume
Reorder
Read Only
Redefine
Reroute
Mirror Geom
Insert Mode
Done

Copy

This command is used to make a single duplication of an existing part feature.

Create

This command is used to create one of the many feature types for Pro/ENGINEER part models.

Delete

This command is used to remove an existing part feature.

Del Pattern

This command is used to remove a pattern of existing part features, while maintaining the pattern's parent feature.

Group

This command is used to associate multiple features so they may be manipulated as a single entity.

Insert Mode

This command is used to place a new feature anywhere in the parent-child relationships of existing geometry. You will be prompted to select the feature that the new feature will be placed *after*. All the features which come after the new feature will be automatically suppressed. To cancel the effects of the **Insert Mode** command, you must either **resume all the suppressed features, or choose Cancel** from the INSERT sub-menu.

Mirror Geom

This command is used to mirror all the existing part geometry about a planar surface or datum plane.

Pattern

This command is used to make multiple duplications of an existing part feature.

Redefine

This command displays a sub-menu of options that allows you to modify *how* a feature was originally created.

Reorder

This command is used to change the order in which features are regenerated. You may not reorder a feature so that it is regenerated before any of its parents.

Reroute

This command is used to change the existing parent-child relationships that were established when the selected feature was created. You are allowed to select new sketching, placement and dimensioning reference features.

Resume

This command is used to **un-suppress** features that have been temporarily hidden using the **Suppress** command.

Suppress

This command is used to *temporarily* remove features from a part. Most commonly used to display a model in a simplified manner, to decrease the regeneration time requirements. Children of the **suppressed** feature will not be recalculated until the parent is resumed.

PART ➡ FEATURE ➡ CREATE

PART
Feature
Modify
Regenerate
Relations
Family Tab
Interchange
Declare
Info
Interface
Set Up
Ref Dim
X-Section
Layer
Program

FEAT
Create
Pattern
Copy
Delete
Del Patern
Group
Supress
Resume
Reorder
Read Only
Redefine
Reroute
Mirror Geom
Insert Mode
Done

FEAT CLASS
Solid
Surface
Datum
Cosmetic
User Defined
Done/Return

Cosmetic

Cosmetic features are those that are *drawn* on a part. They might include stamped logos, or icons like the brightness and contrast icons found on many computer monitor bezels.

Datum

This command opens the Datum menus which are used to add construction aides to a model. See the Datums section of this book.

Solid

This is the most common type of feature in Pro/ENGINEER. This type of feature visually resembles a surface feature, but it differs in that Pro/ENGINEER knows which side of the surrounding surfaces do and/or do not contain material. With this knowledge, Pro/ENGINEER can perform calculations about solid features such as mass properties, center of gravity and others.

Surface

This type of feature is created when itis too difficult to create the necessary geometry using the solid feature creation methods. Once several surface features exist in a model (through creation or by importing them from other CAD packages) you can *stitch* the surfaces together to define a solid feature.

User Defined

This is a special type of feature that many CAD systems refer to as a block or a cell. This feature type consists of many other *sub-features* which have been pre-created and grouped together for re-use. A user defined feature may look exactly like it's original generation, or it may be modified to have different dimensional values.

Feature - Cosmetic

PART ➡ FEATURE ➡ CREATE ➡ COSMETIC

FEAT
- Create
- Pattern
- Copy
- Delete
- Del Patern
- Group
- Supress
- Resume
- Reorder
- Read Only
- Redefine
- Reroute
- Mirror Geom
- Insert Mode
- Done

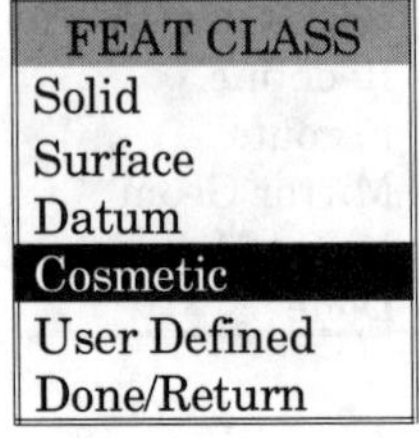

COSMETIC
- Sketch
- Thread
- Groove
- Areas

Groove

The groove feature is used most commonly in the manufacturing process, where a tool is instructed to follow a groove path.

Thread

The thread feature is used to represent the diameter of the threads on a threaded part (like a bolt).

PART ➡ FEATURE ➡ CREATE ➡ COSMETIC ➡ SKETCH

FEAT CLASS
Solid
Surface
Datum
Cosmetic
User Defined
Done/Return

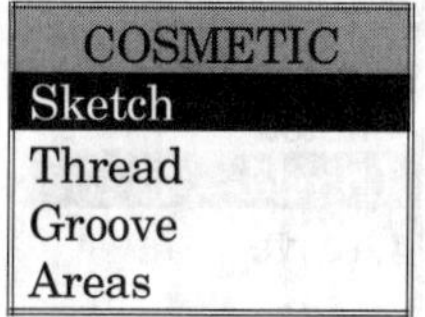

PROJ SECT
Regular Sec
Project Sec
Single
Dim Pattern
Ref Pattern
Xhatch
No Xhatch
Done

Dim Pattern

Creates a pattern of cosmetic features using specified dimensional values to control the pattern.

No X-hatch

Instructs Pro/ENGINEER to create the cosmetic feature without cross-hatching.

Project Sec

This type of cosmetic feature is sketched on a plane and projected onto a single part surface. It may not cross over to other surfaces be cross-hatched or patterned.

Ref Pattern

Creates a pattern of cosmetic features that are dependent upon an existing **Dim Pattern** for their dimensional values.

Regular Sec

This type of cosmetic feature stays exactly where it was sketched. It is a flat feature that lies on a datum, plane surface, or in space.

Single

Creates a single cosmetic feature using the specified options.

X-hatch

Instructs Pro/ENGINEER to create the cosmetic feature with cross-hatching. This command is not available for cosmetic features created using the **Projected Section** option.

PART ➥ FEATURE ➥ CREATE ➥ COSMETIC ➥ THREAD

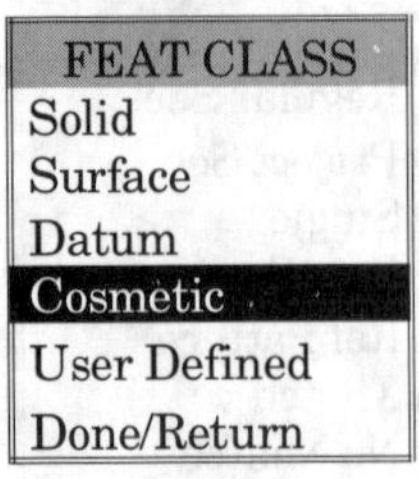

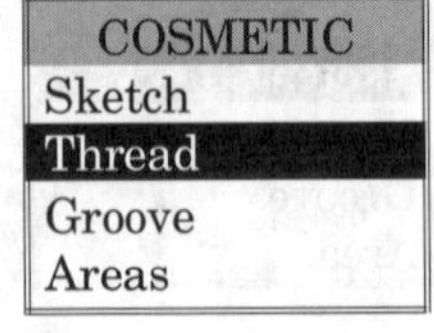

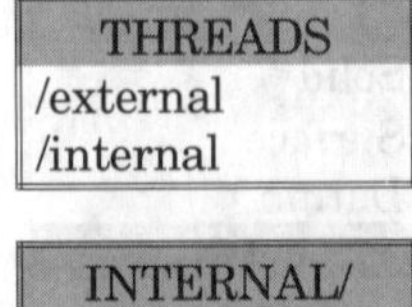

INTERNAL/ EXTERNAL
blind
thru
UP..

/external

This command instructs Pro/ENGINEER that the value entered for the thread diameter is a **minor** value.

/internal

This command instructs Pro/ENGINEER that the value entered for the thread diameter is a **major** value.

blind

This command instructs Pro/ENGINEER that the thread feature's depth will be determined by a user specified value (via the keyboard).

thru

This command instructs Pro/ENGINEER that the thread feature's depth will be determined by the user selecting a starting plane, and an ending edge on the part.

Feature - Solid

PART ➡ FEATURE ➡ CREATE ➡ SOLID

FEAT
Create
Pattern
Copy
Delete
Del Patern
Group
Supress
Resume
Reorder
Read Only
Redefine
Reroute
Mirror Geom
Insert Mode
Done

FEAT CLASS
Solid
Surface
Datum
Cosmetic
User Defined
Done/Return

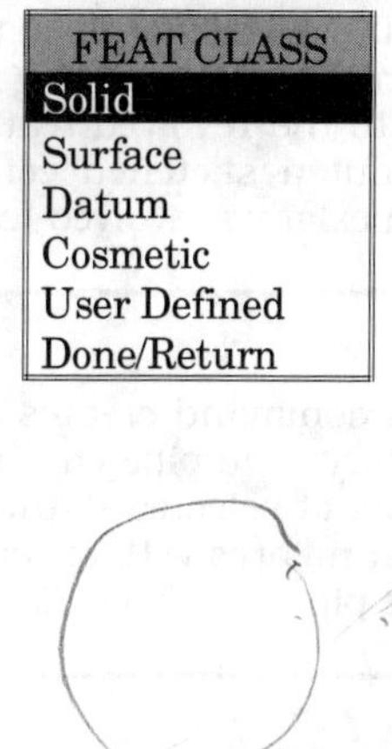

FEATURES
Hole
Shaft
Round
Chamfer
Slot
Cut
Protrusion
Neck
Flange
Rib
Shell
Pipe
Tweak
Intersect

Chamfer

Chamfers remove a flat section of material from the edge or corner of a part. The two types of chamfers (edge and corner) differ in their dimensioning schemes and creation methods.

Cut

Used to remove material. Create a sketched section and define the characteristics of direction (one or both sides) and depth (blind, thru all, etc.). Pro/ENGINEER will prompt you (via a red arrow) to define if material should be removed from the exterior or interior of the section. Use the **Flip** or **Okay** options to change.

Flange

A flange is a special type of feature used to add material around a revolved feature. It is created by sketching a cross-section, aligning both ends of the sketch to the original revolved feature, and revolving the section about a sketched centerline. A flange feature adds material to an existing revolved feature.

Hole

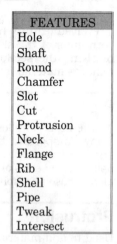

Holes are cylindrical features that remove material. The type of hole generated depends upon the placement characteristics. **Blind** (counterbored and countersunk) holes are created by revolving a sketched section about a centerline. **Through** holes are created by extruding a circular section of a specified size.

Neck

A neck is a special type of revolved slot used to create grooves around a revolved feature. It is created by sketching a cross-section, aligning both ends of the sketch to the revolved feature being trimmed, and revolving the section about a sketched centerline. A neck feature removes material from an existing revolved feature.

Pipe

As the name implies this command creates a circular feature which follows a specified trajectory. The pipe can be either solid or hollow, and the trajectory consists of selected datum points connected by a spline, or a series of lines and arcs with specified bend radii. You may also choose to display the pipe as a 3D centerline.

Protrusion

Used to add material. Create a sketched section and define the characteristics of direction (one or both sides) and depth (blind, thru all, etc.). Pro/ENGINEER will prompt you (via a red arrow) to define if material should be added from the exterior or interior of the section. Use the **Flip** or **Okay** options to change.

Rib

A rib is a special type of feature resulting in a thin *fin* or blade, that is attached to an existing part. It is always sketched from a side view of the part, and grows symmetrically (both sides) about the sketching plane. Because ribs are always attached to existing geometry, they are always drawn as *open* sections with the endpoints aligned.

Round

Rounds are similar to fillets in two-dimensional sketching practices. You identify the common edge between two surfaces and Pro/ENGINEER creates a circular transition along the edge where the two surfaces intersect. There are four types of edges you may create. The most common type (used 99 percent of the time) is called an *edge* round.

Shaft

Shafts are cylindrical features that add material. Shafts are created by revolving a sketched section about a centerline.

Shell

Used to remove material. Select a surface or surfaces of a solid, and this command will remove the surfaces and *hollow out* the solid element, leaving a specified wall thickness. This command is most often used when creating plastic parts.

Slot

Used to remove material. Create a sketched section, and define the characteristics of direction (one or both sides) and depth (blind, thru all, etc.). Pro/ENGINEER will automatically remove the material **enclosed** by the section.

Tweak

This option provides access to many surface modification commands, like draft, local push, and ear.

Feature - Chamfer

PART ➥ FEATURE ➥ CREATE ➥ SOLID ➥ CHAMFER

FEAT CLASS
Solid
Surface
Datum
Cosmetic
User Defined
Done/Return

FEATURES
Hole
Shaft
Round
Chamfer
Slot
Cut
Protrusion
Neck
Flange
Rib
Shell
Pipe
Tweak
Intersect

CHAMF
Edge
Corner
Done/Return

Corner

Instructs Pro/ENGINEER to create a chamfer at the corner of an existing part.

Edge

Instructs Pro/ENGINEER to create a chamfer along the edge of an existing part.

PART ➡ FEATURE ➡ CREATE ➡ SOLID ➡ CHAMFER ➡ CORNER

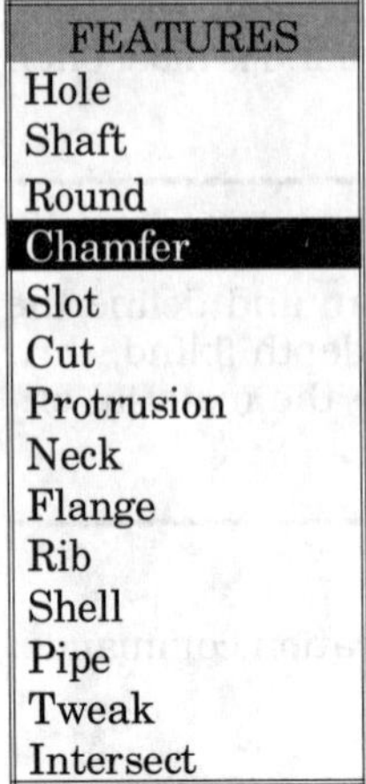

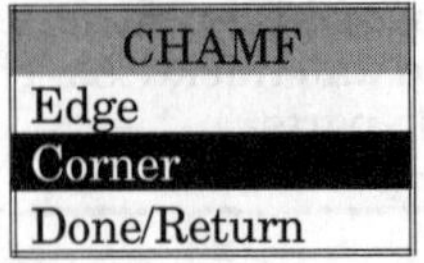

PICK/ENTER
Pick Point
Enter-input

Enter Input

As the three surfaces that define the corner of the part are highlighted, one after the other, you enter (via the keyboard) a distance from the corner for that surface to the trimmed.

Pick Point

As the three surfaces that define the corner of the part are highlighted, one after the other, you pick a point (using the mouse) that specifies a distance from the corner for that surface to the trimmed.

PART ➡ FEATURE ➡ CREATE ➡ SOLID ➡ CHAMFER ➡ EDGE

FEATURES
Hole
Shaft
Round
Chamfer
Slot
Cut
Protrusion
Neck
Flange
Rib
Shell
Pipe
Tweak
Intersect

CHAMF
Edge
Corner
Done/Return

CHAMFERS
45 x d
d x d
d1 x d2
Ang x d
Quit

45 x d

This option may only be used to create a chamfer between two surfaces that are 90 degrees to each other (perpendicular). This is a dimensioning scheme that specifies a common trimming distance, from the edge of the two surfaces, and an angle of 45 degrees from each.

Ang x d

This is a dimensioning scheme which specifies a trimming distance for one surface from the edge of the two surfaces, and an angle from the same.

d x d

This is a dimensioning scheme which specifies a common trimming distance from the edge of the two surfaces.

d1 x d2

This is a dimensioning scheme which specifies a unique trimming distance for each surface involved, from the edge of the two surfaces.

Feature - Cut

PART ➡ FEATURE ➡ CREATE ➡ SOLID ➡ CUTS

FEATURES
Hole
Shaft
Round
Chamfer
Slot
Cut
Protrusion
Neck
Flange
Rib
Shell
Pipe
Tweak
Intersect

FORM
Extrude
Revolve
Sweep
Blend
Use Surfs
Solid
Thin
Done
Quit

THIN OPT
Flip
Okay
Both

NOTE: The Form menu provides commands for creating cuts, slots and protrusion. See entries for Protrusions later in this section for descriptions of Form menu commands.

Flip

Changes the directional arrow to point in the opposite direction. This command must be followed by the **Okay** option, before any material will be removed.

Okay

Accepts the current orientation of the directional arrow for removing material.

Feature - Hole

PART ➡ FEATURE ➡ CREATE ➡ SOLID ➡ HOLE

FEATURES
Hole
Shaft
Round
Chamfer
Slot
Cut
Protrusion
Neck
Flange
Rib
Shell
Pipe
Tweak
Intersect

OPTIONS
Single
Dim Pattern
Ref Pattern
linear
Radial
Coaxial
On Point
Done
Quit

HOLE FORM
Straight
Sketch

SIDES
One Side
Both Sides

SPEC TO
Blind
Thru Next
Thru All
Thru Until
UpTo Pnt/Vtx
UpTo Curve
UpTo Surface
Done
Quit

Blind

This is one of the *depth options* that control how much material is removed. Using this command, you will be prompted to specify (via the keyboard) a value for the feature's depth.

Both Sides

Instructs Pro/ENGINEER that the sketched section will be projected in both directions from the sketching plane. The length of the projection will be determined by the specified *depth option*. Using this option, you may specify a unique depth option for each side.

Coaxial

Creates a hole or shaft that is coaxial to an existing axis (or part or datum). Using this option, only the diameter dimension is required.

Dim Pattern

Creates a pattern of holes or shafts using specified dimensional values to control the pattern.

From To

This is one of the *depth options* that control how much material is removed. Using this command, you will be prompted to specify which existing surfaces will be used to terminate the new feature. The feature will be created between the specified surfaces.

Linear

Instructs Pro/ENGINEER that the dimensioning scheme used to locate a section will refer to linear values offset from two part edges, planar surfaces, or axes.

On Point

Instructs Pro/ENGINEER to place the center of a hole or shaft directly on a selected datum point. The feature will be oriented normal to the surface upon which the point is located.

One Side

Instructs Pro/ENGINEER that a sketched section will be projected in only one direction from the sketching plane. The length of the projection will be determined by the specified *depth option.*

Radial

Instructs Pro/ENGINEER that the dimensioning scheme used to locate a section will refer to polar values.

Ref Pattern

Creates a pattern of holes or shafts which are dependent upon an existing **Dim Pattern** for their dimensional values.

Single

Creates a single hole or shaft using the specified options.

Thru All

This is one of the *depth options* that control how much material is removed. Using this command, the feature will be terminated when it reaches the last surface it encounters.

Thru Next

This is one of the *depth options* that control how much material is removed. Using this command, the feature will be terminated when it reaches the first surface it encounters.

Thru Until

This is one of the *depth options* that control how much material is removed. Using this command, you will be prompted to specify which existing surface will be used to terminate the new feature.

Feature - Pipe

PART ➡ FEATURE ➡ CREATE ➡ SOLID ➡ PIPE

FEATURES
Hole
Shaft
Round
Chamfer
Slot
Cut
Protrusion
Neck
Flange
Rib
Shell
Pipe
Tweak
Intersect

OPTIONS
Geometry
No Geometry
Hollow
Solid
Constant Rad
Multiple Rad
Done
Quit

PROCESS
Done
Quit

CONNECT TYPE
Line/Arc
Spline
Single Point
Whole Array

Constant Rad

This command works with the **Line/Arc** point-connection method and instructs Pro/ENGINEER to create each arc segment in the pipe with a common bend radius.

Geometry

This command instructs Pro/ENGINEER to create the pipe geometry, in addition to storing the information about the pipe feature.

Hollow

This command instructs Pro/ENGINEER to display the pipe feature as hollow, using a specified wall thickness. When used with **no geometry**, this command just stores the value for wall thickness.

Line / Arc

This command creates a network of line segments and arcs between the selected datum points to define the trajectory of the pipe feature.

The radii of the arcs are determined by the **Constant** or **Variable** options.

Multiple Rad

This command works with the **Line/Arc** point-connection method and instructs Pro/ENGINEER to create each arc segment in the pipe with a user-specified bend radius. They may vary as desired.

No Geometry

This command instructs Pro/ENGINEER to display the pipe feature as a 3D centerline and to store information about the pipe feature, but not to create elements to represent the pipe itself.

Solid

This command instructs Pro/ENGINEER to display the pipe feature as a solid tube.

Single Point

This command is used to manually select specific datum points in a datum point array. The selected points will be used with the **Line/Arc** or **Spline** command to create the pipe's trajectory.

Spline

This command creates a 3D spline element between the selected datum points to define the trajectory of the pipe feature.

Whole Array

This command is used to automatically select all the datum points in a datum point array. The selected points will be used with the **Line/Arc** or **Spline** command to create the pipe's trajectory.

Feature - Protrusions

PART ➥ FEATURE ➥ CREATE ➥ SOLID ➥ PROTRUSION

FEATURES
Hole
Shaft
Round
Chamfer
Slot
Cut
Protrusion
Neck
Flange
Rib
Shell
Pipe
Tweak
Intersect

FORM
Extrude
Revolve
Sweep
Blend
Use Surfs
Solid
Thin
Done
Quit

Blend

A blend is a feature consisting of at least two planar sections that are joined together at their edges, using *transitional* surfaces, to construct a single feature. Another type of blend, called a *parallel* blend, is created from a single section containing multiple contours, called *sub-sections*.

Extrude

An extrusion is a feature created by projecting a sketched section normal to a specified sketching plane. The sketching plane can be selected, or created using the **Make Plane** option.

Revolve

This type of feature is created by revolving a sketched section about a centerline. The sweep angle is specified from the ANGLE sub-menu. When creating the section, the first centerline created will act as the axis of revolution. The feature's section must be completely constructed on one side of the centerline and be closed.

Sweep

This type of feature is created by sketching or selecting a path (trajectory), then sketching a section which will follow *along* the specified trajectory.

Thin

Used to sketch a section and apply a thickness as it is extruded, revolved, swept, or blended.

Use Srfs

Allows you to transform **surface features** (datum surfaces and quilts) into construction features.

PART ➥ FEATURE ➥ CREATE ➥ SOLID ➥ PROTRUSION ➥ BLEND

FEATURES
Hole
Shaft
Round
Chamfer
Slot
Cut
Protrusion
Neck
Flange
Rib
Shell
Pipe
Tweak
Intersect

FORM
Extrude
Revolve
Sweep
Blend
Use Surfs
Solid
Thin
Done
Quit

OPTIONS
Parallel
Rotational
General
Open
Closed
Straight
Smooth
Blind
Thru Next
Thru All
Thru Until
From To
Regular Sec
Project Sec
Single
Dim Pattern
Ref Pattern
Done
Quit

Blind

This is one of the *depth options* that control how long the new feature will be. Using this command, you will be prompted to specify (via the keyboard) a value for the feature's depth.

Closed

This command is applicable when creating a blend between *non parallel* sections. Using this option, the first section selected will also automatically be used as the last section of the blend.

Dim Pattern

Creates a pattern of blends, using specified dimensional values to control the pattern.

From To

This is one of the *depth options* that control the length of the new feature. Using this command, you will be prompted to specify which existing surfaces will be used to terminate the new feature. The feature will be created between the specified surfaces.

General

Creates a blend using either the **No Profile** or **One Profile** options.

Open

This command is applicable when creating a blend between *non parallel* sections. Using this option, the two selected sections will act as the blend's beginning and ending points, creating an **open** blend.

Parallel

This option informs Pro/ENGINEER that all blend sections lie on parallel planes. This includes projected section blends, also.

Project Sec

This is a special type of parallel blend that allows you to sketch a section on a planar surface and project that section onto any two surfaces. This type of blend keeps its exact dimensions as shown on the sketch plane (no distortion).

Ref Pattern

Creates a pattern of blends that are dependent upon an existing *Dim Pattern* for their dimensional values.

Rotational

This option informs Pro/ENGINEER that the blend sections are rotated about the **Y-axis** by up to a maximum total value of 240 degrees.

Single

Creates a single blend, using the specified options.

Smooth

When this option is chosen, the transitional surfaces of the blend are created using spline surfaces.

Straight

When this option is chosen, transitional surfaces of the blend are created using ruled surfaces.

Thru All

This is one of the *depth options* that control the length of the new feature. Using this command, the feature will be terminated when it reaches the last surface it encounters.

Thru Next

This is one of the *depth options* that control the length of the new feature. Using this command, the feature will be terminated when it reaches the first surface it encounters.

Thru Until

This is one of the *depth options* that control the length of the new feature. Using this command, you will be prompted to specify which existing surface will be used to terminate the new feature.

PART ➡ FEATURE ➡ CREATE ➡ SOLID ➡ PROTRUSION ➡ BLEND ➡ DONE

FORM
Extrude
Revolve
Sweep
Blend
Use Surfs
Solid
Thin
Done
Quit

OPTIONS
Parallel
Rotational
General
Open
Closed
Straight
Smooth
Blind
Thru Next
Thru All
Thru Until
From To
Regular Sec
Project Sec
Single
Dim Pattern
Ref Pattern
Done
Quit

CAP TYPE
Smooth
Sharp

Sharp

Allows the blend geometry of a non-parallel blend to flow straight towards the point section.

Smooth

Forces the blend geometry of a non-parallel blend to be tangent to the point section.

PART ➥ FEATURE ➥ CREATE ➥ SOLID ➥ PROTRUSION ➥ BLEND ➥ GENERAL or SMOOTH

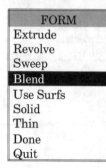

FORM
Extrude
Revolve
Sweep
Blend
Use Surfs
Solid
Thin
Done
Quit

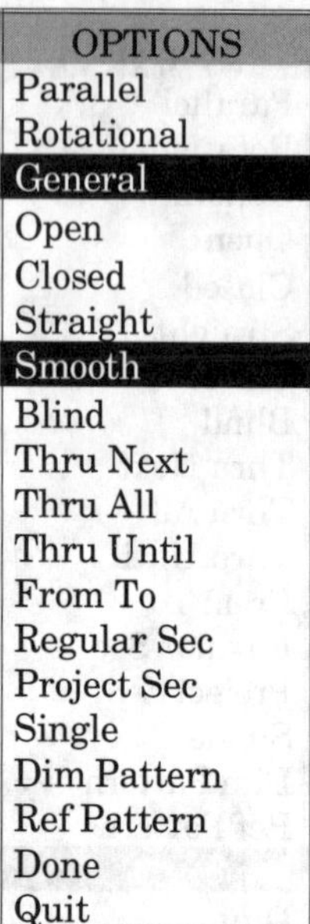

TANGENCY OPT
No Opt Tan
Optional Tan
Done
Quit

NOTE: Both the General and Smooth blend options open the TANGENCY OPT menu.

No Opt Tan

Instructs Pro/ENGINEER that the resulting blend will not necessarily be tangent to any existing surfaces.

Optional Tan

Allows you to specify surfaces to which the resulting swept blend will be tangent.

PART ➡ FEATURE ➡ CREATE ➡ SOLID ➡ PROTRUSION ➡ EXTRUDE

FEATURES
- Hole
- Shaft
- Round
- Chamfer
- Slot
- Cut
- Protrusion
- Neck
- Flange
- Rib
- Shell
- Pipe
- Tweak
- Intersect

FORM
- Extrude
- Revolve
- Sweep
- Blend
- Use Surfs
- Solid
- Thin
- Done
- Quit

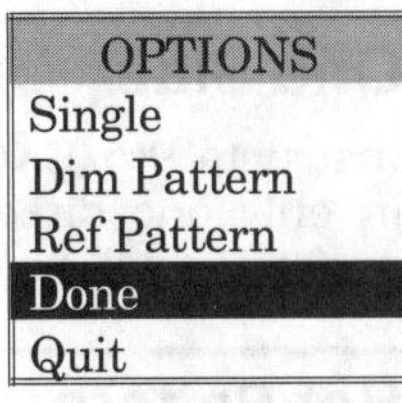

SIDES
- One Side
- Both Sides

SPEC TO
- Blind
- Thru Next
- Thru All
- Thru Until
- UpTo Pnt/Vtx
- UpTo Curve
- UpTo Surface
- Done
- Quit

Blind

This is one of the *depth options* that control how long the new feature will be extruded, or how much material will be removed. The type of function (add or remove material) is determined by the active command. Using this command, you will be prompted to specify (via the keyboard) a value for the feature's depth.

Both Sides

Instructs Pro/ENGINEER that the sketched section will be projected in both directions from the sketching plane. The length of the projection will be determined by the specified *depth option.* Using this option, you may specify a unique depth option for each side.

Dim Pattern

Creates a pattern of extrusions, using specified dimensional values to control the pattern.

From To

This is one of the *depth options* that control the length of the new feature. Using this command, you will be prompted to specify which existing surfaces will be used to terminate the new feature. The feature will be created between the specified surfaces.

One Side

Instructs Pro/ENGINEER that the sketched section will be projected in only one direction from the sketching plane. The length of the projection will be determined by the specified *depth option.*

Ref Pattern

Creates a pattern of extrusions that are dependent upon an existing **Dim Pattern** for their dimensional values.

Single

Creates a single extrusion using the specified options.

Thru All

This is one of the *depth options* that control the length of the new feature. Using this command, the feature will be terminated when it reaches the last surface it encounters.

Thru Next

This is one of the *depth options* that control the length of the new feature. Using this command, the feature will be terminated when it reaches the first surface it encounters.

Thru Until

This is one of the *depth options* that control the length of the new feature. Using this command, you will be prompted to specify which existing surface will be used to terminate the new feature.

Up to Curve

This is one of the *depth options* that control the length of the new feature. Using this command, you will be prompted to specify a plane which is parallel to the sketching plane, and an edge, axis, or datum curve through which the feature will pass.

Up to Pnt / Vtx

This is one of the *depth options* that control the length of the new feature. Using this command, you will be prompted to specify a plane which is parallel to the sketching plane, and a point or vertex which the feature will pass through.

Up to Surface

This is one of the *depth options* that control the length of the new feature. Using this command, you will be prompted to specify an existing surface. For solid features, the surface can be another part surface which is not planar, or a datum plane which need not be parallel to the sketching plane. For surface features, the selected surface can only be a datum plane which is parallel to the sketching plane.

PART ➡ FEATURE ➡ CREATE ➡ SOLID ➡ PROTRUSION ➡ REVOLVE

FEATURES
Hole
Shaft
Round
Chamfer
Slot
Cut
Protrusion
Neck
Flange
Rib
Shell
Pipe
Tweak
Intersect

FORM
Extrude
Revolve
Sweep
Blend
Use Surfs
Solid
Thin
Done
Quit

OPTIONS
Variable
90
180
270
360
Single
Dim Pattern
Ref Pattern
Done
Quit

90

This command specifies a sweep angle of 90 degrees to use when creating a revolved feature.

180

This command specifies a sweep angle of 180 degrees to use when creating a revolved feature.

270

This command specifies a sweep angle of 270 degrees to use when creating a revolved feature.

360

This command specifies a sweep angle of 360 degrees to use when creating a revolved feature.

Dim Pattern

Creates a pattern of revolved features, using specified dimensional values to control the pattern.

Ref Pattern

Creates a pattern of revolved features that are dependent upon an existing **Dim Pattern** for their dimensional values.

Single

Creates a single revolved feature, using the specified options.

Variable

This command specifies a user defined sweep angle (between 0 and 360) to use when creating a revolved feature.

PART ➡ FEATURE ➡ CREATE ➡ SOLID ➡ PROTRUSION ➡ SWEEP

FEATURES
Hole
Shaft
Round
Chamfer
Slot
Cut
Protrusion
Neck
Flange
Rib
Shell
Pipe
Tweak
Intersect

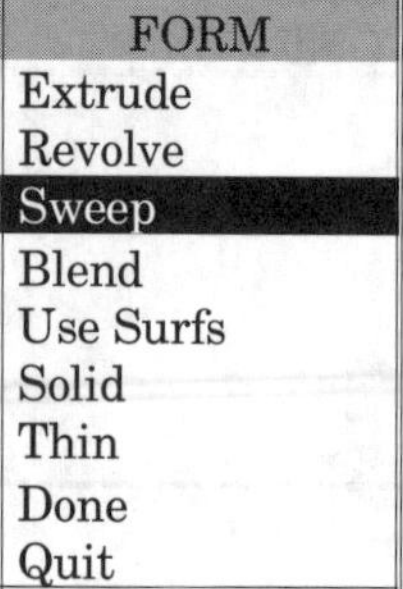

SWEEP TRAJ
Sketch
X-Section
Along Pipe
Along Curve
Add Inn Fcs
No Inn Fcs
Merge Ends
Free Ends
Normal Traj
Normal Surf
One Profile
Multi Prof.
Done
Quit

Add Inn Fcs

Instructs Pro/ENGINEER to **cap** the ends of the resulting sweep by adding top and bottom faces.

Along Curve

Instructs Pro/ENGINEER to use a specified datum curve for a sweep trajectory.

Along Pipe

Instructs Pro/ENGINEER to use a specified pipe centerline for a sweep trajectory.

Free Ends

Instructs Pro/ENGINEER *not* to merge either end of the sweep to intersecting part geometry. Creates the sweep exactly as sketched.

Merge Ends

Instructs Pro/ENGINEER to merge both ends of the sweep to intersecting part geometry.

Multi Prof.

Instructs Pro/ENGINEER the feature has longitudinal curves (in addition to the trajectory) that affect the section's orientation and dimensional values.

No Inn Fcs

Instructs Pro/ENGINEER *not* to **cap** the ends of the resulting sweep.

Normal Surf

Instructs Pro/ENGINEER the sketched section will, if necessary, pivot about the trajectory.

Normal Traj

Instructs Pro/ENGINEER the sketched section will be oriented normal to the trajectory, at a fixed angle.

One Profile

Instructs Pro/ENGINEER the feature has only one trajectory along which the section will be swept.

Sketch

Instructs Pro/ENGINEER to place you in a sketcher session so you may sketch the sweep's trajectory.

X-section

Instructs Pro/ENGINEER to use a specified planar cross-section for a sweep trajectory.

PART ➡ FEATURE ➡ CREATE ➡ SOLID ➡ PROTRUSION ➡ SWEEP ➡ ALONG CURVE or MULTI PROF

FEATURES
Hole
Shaft
Round
Chamfer
Slot
Cut
Protrusion
Neck
Flange
Rib
Shell
Pipe
Tweak
Intersect

SWEEP TRAJ
Sketch
X-Section
Along Pipe
Along Curve
Add Inn Fcs
No Inn Fcs
Merge Ends
Free Ends
Normal Traj
Normal Surf
One Profile
Multi Prof.
Done
Quit

TANGENCY OPT
No Opt Tan
Optional Tan
Done
Quit

Note: Selecting either Along Curve or Multi Prof from the SWEEP TRAJ menu opens the TANGENCY OPT menu.

No Opt Tan

Instructs Pro/ENGINEER that the resulting sweep will not necessarily be tangent to any existing surfaces.

Optional Tan

Allows you to specify surfaces to which the resulting sweep will be tangent.

PART ➡ FEATURE ➡ CREATE ➡ SOLID ➡ PROTRUSION ➡ THIN

FEATURES
Hole
Shaft
Round
Chamfer
Slot
Cut
Protrusion
Neck
Flange
Rib
Shell
Pipe
Tweak
Intersect

FORM
Extrude
Revolve
Sweep
Blend
Use Surfs
Solid
Thin
Done
Quit

THIN OPT
Flip
Okay
Both

Both

Instructs Pro/ENGINEER that the sketched section will be projected in both directions from the sketching plane. The length of the projection will be determined by the specified *depth option.* Using this option, you may specify a unique depth option for each side.

Flip

Changes the directional arrow to point in the opposite direction. This command must be followed by the **Okay** option before any material will be removed.

Okay

Accepts the current orientation of the directional arrow for removing material.

Feature - Round

PART ➥ FEATURE ➥ CREATE ➥ SOLID ➥ ROUND

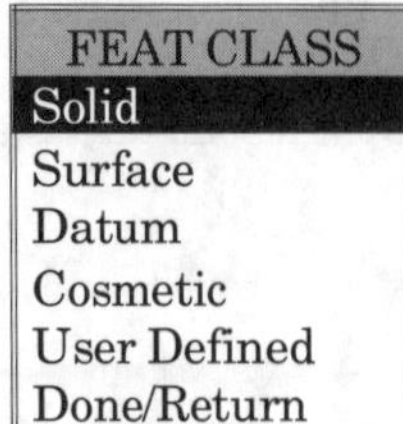

FEATURES
Hole
Shaft
Round
Chamfer
Slot
Cut
Protrusion
Neck
Flange
Rib
Shell
Pipe
Tweak
Intersect

ROUNDS
Edge
Edge-Surf
Remove Surf
Surf-Surf
Norm to Edge
Constant
Multi Const
Variable
Conic
Circular
Done
Quit

Circular

Instructs Pro/ENGINEER to create a round along an edge, using a circular cross-section to define the radius of the round.

Conic

Instructs Pro/ENGINEER to create a round along an edge, using a conic cross-section to define the radius of the round. This option allows the cross-section to have an extra conic parameter to control its shape.

Constant

This option is one method for controlling the radii of a round. Using this option, you will be prompted only once for the radius value. As a result, the round will have a constant radius value.

Edge

This command creates a round along a selected edge, the most common type of round.

Edge — Surf

This type of round is used to create a circular transition surface between an edge and tangent to a selected surface.

Multi — Const

This option is one method for controlling the radii of a round. This option works with the edge *selection* option commands, and allows you to enter different values in accordance with the selection method. For example, if you used **Chain** as the selection option for one section of the round, and **Loop** for another section of the round, using this option, you could enter different values for the two sections identified.

Remove Surf

This type of round is used to completely **replace** a surface between two selected edges. The radius of the new rounded surface is determined by the edges and surfaces adjacent to the surface being replaced.

Surf — Surf

This type of round is used to create a circular transition surface between surfaces which may or may not share a common edge. This round always has a *constant* radius.

Variable

This option is one method for controlling the radii of a round. Using this option, you will be prompted for the radius value at every edge vertex and datum point. As a result, the round can have a different radius value at each vertex.

PART ➥ FEATURE ➥ CREATE ➥ SOLID ➥ ROUND ➥ EDGE

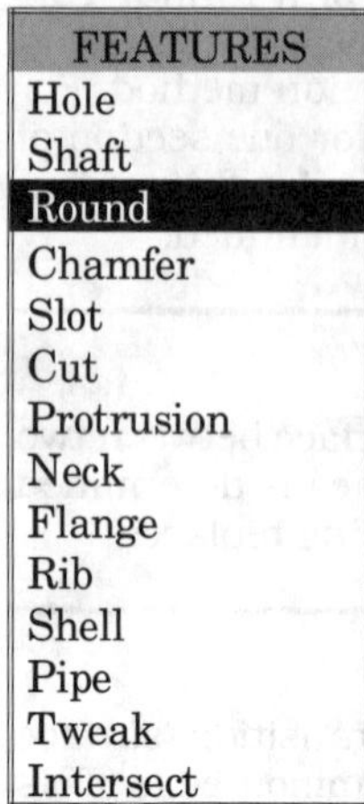

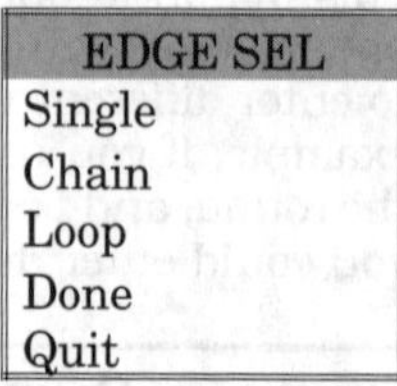

Chain

This command is used to select all the edges required to form a continuous chain. The edges must be tangent at their endpoints. When this command is used with the **Variable** option, you will only be prompted for a radius value at the beginning and end points of the chain.

Loop

This command is used to select a surface. The edges that bound the surface will be rounded using the specified options. If more than one boundary exists, you can choose which edges to round. You will be prompted for a radius value at the endpoints of every edge used in the loop.

Single

This command is used to select one edge to be rounded at a time, using the specified options.

Feature - Shaft

PART ➡ FEATURE ➡ CREATE ➡ SOLID ➡ SHAFT

FEATURES
Hole
Shaft
Round
Chamfer
Slot
Cut
Protrusion
Neck
Flange
Rib
Shell
Pipe
Tweak
Intersect

OPTIONS
Single
Dim Pattern
Ref Pattern
Iinear
Radial
Coaxial
On Point
Done
Quit

HOLE FORM
Straight
Sketch

SIDES
One Side
Both Sides

SPEC TO
Blind
Thru Next
Thru All
Thru Until
UpTo Pnt/Vtx
UpTo Curve
UpTo Surface
Done
Quit

NOTE: Commands and options used to create shafts are the same as those used to create holes. For descriptions of these menu items, see the entry for holes earlier in this section.

Feature - Slot

PART ➡ FEATURE ➡ CREATE ➡ SOLID ➡ SLOTS

FEATURES
Hole
Shaft
Round
Chamfer
Slot
Cut
Protrusion
Neck
Flange
Rib
Shell
Pipe
Tweak
Intersect

FORM
Extrude
Revolve
Sweep
Blend
Use Surfs
Solid
Thin
Done
Quit

THIN OPT
Flip
Okay
Both

NOTE: The Form menu provides commands for creating cuts, slots and protrusion. See entries for Protrusions earlier in this section for descriptions of Form menu commands.

Flip

Changes the directional arrow to point in the opposite direction. This command must be followed by the **Okay** option, before any material will be removed.

Okay

Accepts the current orientation of the directional arrow for removing material.

Feature - Tweak

PART ➡ FEATURE ➡ CREATE ➡ SOLID ➡ TWEAK

FEAT CLASS
Solid
Surface
Datum
Cosmetic
User Defined
Done/Return

FEATURES
Hole
Shaft
Round
Chamfer
Slot
Cut
Protrusion
Neck
Flange
Rib
Shell
Pipe
Tweak
Intersect

TWEAK
Draft
Local Push
Radius Dome
Section Dome
Offset
Replace
Ear
Lip
Patch

Draft

This command adds a draft angle (15 degrees positive or negative) to existing cylindrical or planar surfaces. Draft angles can be added to single surfaces, a group, or planar surfaces.

Ear

This command creates a protrusion that is extruded along the top of a surface and is *bent* at the base. The angle of the bend may either be a default value of 90 degrees or user specified.

Lip

Although this command works with parts that are intended to be assembled, it is *not* an assembly feature and must be created on each part separately. A lip feature is intended to be created on mating surfaces of two parts. The feature is created as a protrusion on one part and as a cut on the other.

Local Push

This command deforms an area of an existing surface. The area is determined by sketching a circular or rectangular section. All **local push** features are created with a default height value. Once completed, you may edit the height for a more dramatic push. A negative height value pushes *into* the surface, leaving a depression.

Offset

This command deforms either an area of an existing surface or the entire surface. The area method is determined by sketching a section. A negative offset value removes material from the surface.

Patch

This command replaces an area of an existing surface or surface(s) with a surface feature.

Radius Dome

This command deforms an existing surface to create a dome. The dome is determined by a user specified dome radius and an offset value. A negative radius value *pushes* into the surface definition, creating a concave surface.

Replace

This command replaces an existing surface with a surface feature or a datum plane.

Section Dome

This command deforms an existing surface to create a dome. The dome is determined by a sweep or blend of multiple cross-sections.

PART ➥ FEATURE ➥ CREATE ➥ SOLID ➥ TWEAK ➥ DRAFT

FEATURES
Hole
Shaft
Round
Chamfer
Slot
Cut
Protrusion
Neck
Flange
Rib
Shell
Pipe
Tweak
Intersect

TWEAK
Draft
Local Push
Radius Dome
Section Dome
Offset
Replace
Ear
Lip
Patch

DRAFT OPTION
Regular
Split
Curve Driven
Constant
Variable
Neutral
Ref & Neut
Unmirrored
Mirrored
Done
Quit

Constant

This command adds a constant draft angle along a drafted surface.

Curve Driven

This command creates a draft using a selected datum curve as the *parting* line. Using this command, you may specify different draft angles for each side of the datum curve.

Mirrored

This command creates a draft that is mirrored about the neutral plane. Using this command, you may either remove or add material to each side of the neutral plane.

Neutral

This command instructs Pro/ENGINEER that the neutral plane and the reference plane (used for calculating the draft angles) are one and the same.

Ref & Neutral

This command instructs Pro/ENGINEER that the neutral plane and the reference plane (used for calculating the draft angles) are different elements and will be selected separately. This option should be chosen when the neutral and reference planes are not parallel, or when creating variable (including mirrored) drafts.

Regular

This command adds a taper of a specific angle to a **whole** surface. Any other type of draft requires Pro/FEATURE.

Split

This command adds different draft angles to different portions of a selected surface.

Unmirrored

This command creates a draft about the neutral plane that is not mirrored, and removes material from one side of the neutral plane while adding material to the other.

Variable

This command adds different draft angles at various points along a drafted surface.

PART ➥ FEATURE ➥ CREATE ➥ SOLID ➥ TWEAK ➥ EAR

FEATURES
Hole
Shaft
Round
Chamfer
Slot
Cut
Protrusion
Neck
Flange
Rib
Shell
Pipe
Tweak
Intersect

TWEAK
Draft
Local Push
Radius Dome
Section Dome
Offset
Replace
Ear
Lip
Patch

EAR
Variable
90 deg tab
Single
Pattern
Done
Quit

90 deg tan

Creates an ear feature that is bent at a 90 degree angle. This command does not create a modifiable angle dimension for the angle between the ear and the surface that the ear is extruded from.

Pattern

Creates a pattern of ear features, using specified dimensional values to control the pattern.

Single

Creates a single ear feature, using the specified options.

Variable

Creates an ear feature that is bent at a user defined angle. This command creates a modifiable angle dimension for the angle between the ear and the surface that the ear is extruded from.

PART ➡ FEATURE ➡ CREATE ➡ SOLID ➡ TWEAK ➡ LOCAL PUSH

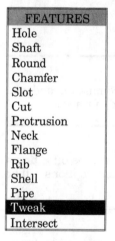

FEATURES
Hole
Shaft
Round
Chamfer
Slot
Cut
Protrusion
Neck
Flange
Rib
Shell
Pipe
Tweak
Intersect

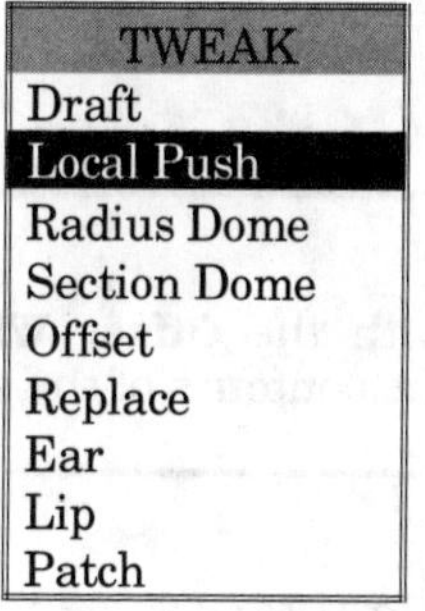

POSITION
Single
Dim Pattern
Ref Pattern

Dim Pattern

Creates a pattern of local push features, using specified dimensional values to control the pattern.

Ref Pattern

Creates a pattern of local push features that are dependent upon an existing **Dim Pattern** for their dimensional values.

Single

Creates a single local push feature, using the specified options.

PART ➥ FEATURE ➥ CREATE ➥ SOLID ➥ TWEAK ➥ OFFSET

FEATURES
Hole
Shaft
Round
Chamfer
Slot
Cut
Protrusion
Neck
Flange
Rib
Shell
Pipe
Tweak
Intersect

TWEAK
Draft
Local Push
Radius Dome
Section Dome
Offset
Replace
Ear
Lip
Patch

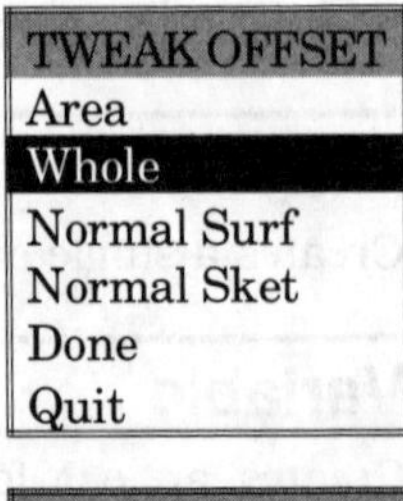

WHOLE OPTION
Pick Loops
All Loops
Done
Quit

All Loops

This command works with the **Offset Whole** command, and will automatically select all the contours of the selected surface.

Area

This command instructs Pro/ENGINEER that you will be offsetting a sketched area of a surface. The appropriate sub-menu options will be displayed after choosing this command.

Normal Sket

This command works with the **Offset Area** command, and instructs Pro/ENGINEER to create the resulting offset normal to the sketching plane.

Normal Surf

This command works with the **Offset Area** command, and instructs Pro/ENGINEER to create the resulting offset normal to the selected surface.

Pick Loops

This command works with the **Offset Whole** command, and will highlight each contour of the selected surface in turn. You must use the selection commands accept, next, or **previous** to manually select the contours you desire.

Whole

This command instructs Pro/ENGINEER that you will be offsetting an entire surface. The appropriate sub-menu options will be displayed after choosing this command.

PART ➥ FEATURE ➥ CREATE ➥ SOLID ➥ TWEAK ➥ SECTION DOME

FEATURES
Hole
Shaft
Round
Chamfer
Slot
Cut
Protrusion
Neck
Flange
Rib
Shell
Pipe
Tweak
Intersect

TWEAK
Draft
Local Push
Radius Dome
Section Dome
Offset
Replace
Ear
Lip
Patch

SECTION DOME
Sweep
Blend
No Profile
One Profile
Done
Quit

Blend

This command creates a section dome by blending two or more sketched sections. You have the choice of using a profile with the sketched sections, if desired.

No Profile

This command creates a section dome by placing transition surfaces between two or more sketched sections. You will be prompted to select a sketching plane and create the two required sections. A section that drops below the surface will remove material, and sections above the surface will add material. This command will not work with the **Sweep** option for creating section domes.

One Profile

This command creates a section dome by sweeping a selected profile with two or more sketched sections. You will be prompted to select a sketching plane that is perpendicular to the dome profile, and to create the two required sections. A section that drops below the surface will remove material, and sections above the surface will add material.

Sweep

This command creates a section dome by sweeping the first selected profile along the second, sweeping the second profile along the first, then using the mathematical average of the two surfaces.

PART ➡ FEATURE ➡ CREATE ➡ SURFACE

FEAT CLASS
Solid
Surface
Datum
Cosmetic
User Defined
Done/Return

SURFACE
New
Merge
Trim
Extend
Transform

Extend

This command is used to create new surface features by extending the boundaries of existing surfaces to a datum plane. You must extend the surface in a direction normal to the datum plane.

Merge

This command is used to merge two surfaces together, resulting in one surface.

New

This command is used to create new surface features. You do so by defining boundaries for the new features, by copying existing surfaces, or through many other techniques.

Transform

This command is used to create new surface features by rotating, mirroring, or translating datum curves and surface features.

Trim

This command is used to *trim* a surface feature. The trimming plane can either be defined by another surface or by the selected surface's own silhouette (a silhouette is the profile of a curved surface as it appears in a view).

Feature - Surface

PART ➥ FEATURE ➥ CREATE ➥ SURFACE ➥ MERGE

FEAT CLASS
Solid
Surface
Datum
Cosmetic
User Defined
Done/Return

SURFACE
New
Merge
Trim
Extend
Transform

SURF MERGE
Join
Intersect
Done

Intersect

This method of merging two surfaces is used when the surfaces eventually intersect each other, and you wish to merge them along that intersection. Using this command, you will be prompted for the portions of the two surfaces that you wish to keep (after the **merge** process is completed).

Join

This method of merging two surfaces is used when the surfaces *share* a common boundary. This command will process much faster than the **intersect** method, due to fewer calculation requirements.

PART ➥ FEATURE ➥ CREATE ➥ SURFACE ➥ NEW

FEAT CLASS
Solid
Surface
Datum
Cosmetic
User Defined
Done/Return

SURFACE
New
Merge
Trim
Extend
Transform

SURF FORM
Extrude
Revolve
Sweep
Blend
Flat
Offset
Copy
Advanced
Done
Quit

Advanced

This command allows you to create surfaces using more complex definitions.

Blend

A blend is a feature consisting of at least two planar sections that are joined together at their edges, using *transitional* surfaces, to construct a single feature. Another type of blend, called a *parallel* blend, is created from a single section containing multiple contours, called *subsections*.

Copy

This command instructs Pro/ENGINEER that you will be creating new surface features by selecting part surfaces to duplicate. The resulting surface feature will be created directly on top of the part surface.

Extrude

An extrusion is a type of surface feature created by projecting a sketched section normal to a specified sketching plane. The sketching plane can be selected, or created using the **Make Plane** option.

Flat

This command is used to create a *planar* surface by sketching its boundaries on a sketching plane. It is similar to the **Extrude** command, but the resulting sketch is not projected.

Offset

This command instructs Pro/ENGINEER that you will be creating a new surface feature that has the same shape as an existing surface, but unlike with the Ccopy command, the resulting surface will be *offset* by a user specified distance.

Revolve

This type of surface feature is created by revolving a sketched section about a centerline. The sweep angle is specified from the ANGLE sub-menu. When creating the section, the first centerline created will act as the axis of revolution. The feature's section must be completely constructed on one side of the centerline, and be closed.

Sweep

This type of surface feature is created by sketching or selecting a path (trajectory), then sketching a section which will be extruded **along** the specified trajectory.

PART ➡ FEATURE ➡ CREATE ➡ SURFACE ➡ NEW ➡ BOUNDARIES

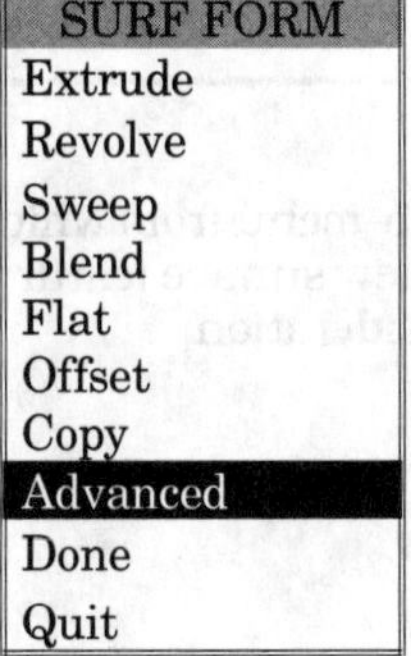

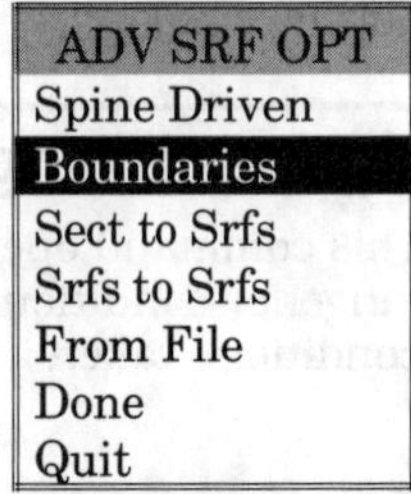

Approx Blend

This command creates a blended surface, using boundary curves and approximation curves as a guide. How closely the surface follows the curves is determined by the number of surface patches created. The higher the number of patches, the closer the surface will follow the curve data.

Blended Surf

This command creates a blended surface, using boundary curves to define it. Using this command, you may also choose additional *interior* curves to further define the form of the surface feature.

Conic Surf

This command creates a conic surface, using two selected boundary elements and a control curve. Composite curves are not allowed. You must also define how the conic curve will react with the control curve, by selecting one of the optional parameters.

No Opt Tan

This command instructs Pro/ENGINEER that the new surface feature will have no tangency conditions, and not to open the tangency options sub-menu.

Optional Tangent

This command opens a sub-menu from which you may choose various tangency conditions. The new surface feature will be created with these conditions taken into consideration.

PART ➡ FEATURE ➡ CREATE ➡ SURFACE ➡ NEW ➡ BOUNDARIES ➡ BLENDED SURF

SURF FORM
Extrude
Revolve
Sweep
Blend
Flat
Offset
Copy
Advanced
Done
Quit

ADV SRF OPT
Spine Driven
Boundaries
Sect to Srfs
Srfs to Srfs
From File
Done
Quit

OPTIONS
Blended Surf
Conic Surf
Approx Blend
Shoulder Crv
Tangent Crv
No Opt Tan
Optional Tan
Done

BLEND TYPE
Arc Length
Pointwise
Done
Quit

Arc Length

This command instructs Pro/ENGINEER to create a surface by dividing the selected boundaries by an equal amount, then blending a surface between the boundary elements (piece by piece).

Pointwise

This command instructs Pro/ENGINEER to create a surface by blending a surface between the boundary elements (point by point). This type of surface creation method does **not** divide the elements equally, and point one of the first boundary element will be connected to point one of the second, and so on.

PART ➡ FEATURE ➡ CREATE ➡ SURFACE ➡ NEW ➡ BOUNDARIES ➡ CONIC SURF

SURF FORM
Extrude
Revolve
Sweep
Blend
Flat
Offset
Copy
Advanced
Done
Quit

ADV SRF OPT
Spine Driven
Boundaries
Sect to Srfs
Srfs to Srfs
From File
Done
Quit

OPTIONS
Blended Surf
Conic Surf
Approx Blend
Shoulder Crv
Tangent Crv
No Opt Tan
Optional Tan
Done

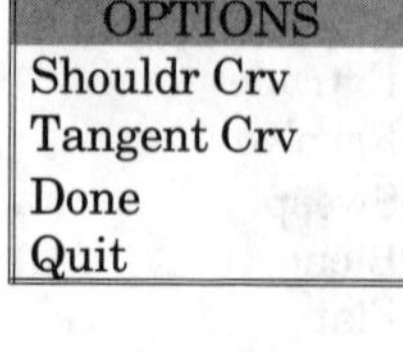

Shouldr Crv

This command instructs Pro/ENGINEER to create a conic surface feature that passes through the control curve. The control curve defines the location of conic shoulders for each cross-section.

Tangent Crv

This command instructs Pro/ENGINEER to create a conic surface feature that does **not** pass through the control curve. The control curve defines the line that passes through the *intersections* of conic cross-sections' asymptotes.

PART ➥ FEATURE ➥ CREATE ➥ SURFACE ➥ NEW ➥ BOUNDARIES ➥ OPTIONAL TANGENT

SURF FORM
- Extrude
- Revolve
- Sweep
- Blend
- Flat
- Offset
- Copy
- **Advanced**
- Done
- Quit

ADV SRF OPT
- Spine Driven
- **Boundaries**
- Sect to Srfs
- Srfs to Srfs
- From File
- Done
- Quit

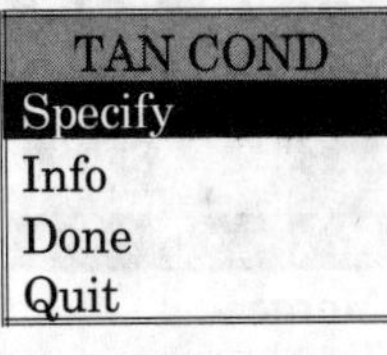

BNDRY COND
- Free Bndry
- Sel Tan Srf
- Tan To Sket
- Nrm To Sket
- Tan To Bndry
- Done Bndry
- Quit Bndry

Free Bndry

This command instructs Pro/ENGINEER to create a blended surface feature with **no** tangency defined.

Nrm to Sket

This command instructs Pro/ENGINEER to create a blended surface feature that is tangent to the *normal of* the curve plane.

Sel Tan Srf

This command instructs Pro/ENGINEER to create a blended surface feature that is tangent to an existing surface.

Tan to Sket

This command instructs Pro/ENGINEER to create a blended surface feature that is tangent to the curve plane.

Tan to Bndry

This command instructs Pro/ENGINEER to create a surface feature that is blended along a boundary tangent to all latitudinal curves. You only see this option when you choose boundary curves in two directions.

PART ➥ FEATURE ➥ CREATE ➥ SURFACE ➥ NEW ➥ COPY ➥ SELECT

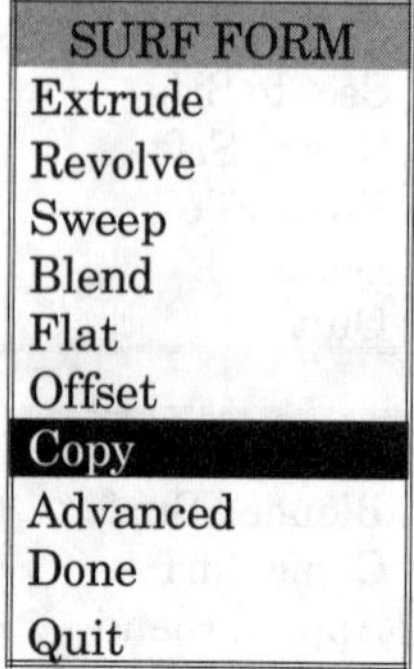

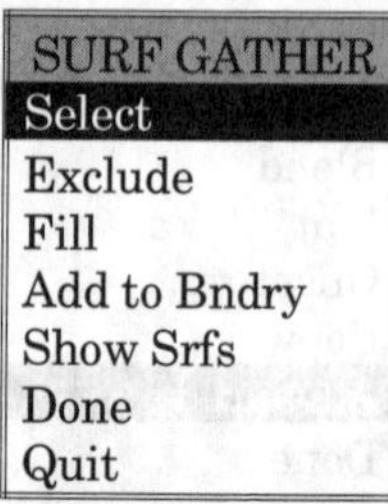

Add to Bndry

This command works with the **Surf & Bnd** selection method, and allows you to add extra surfaces or outer loops to your boundary definition.

Exclude

This command allows you to identify surfaces or outer loops that you want *exlcuded* during the **copy** process. This command is available when using the **Surfaces** selection method.

Fill

This command instructs Pro/ENGINEER to "fill-in" voids (holes, slots), called inner loops, that reside on a selected surface. You may either pick and choose the voids to fill (**Loops**) or you may select the option to fill-all the voids (**All**).

Select

This command allows you to select single surfaces to copy. You may choose this command any time during the **copy** process.

Show Srfs

This command displays the current parameters for gathered surfaces. You may continue to refine your selection parameters prior to completing the **copy** process.

Surfaces

This is one method of selecting which part surfaces will be copied to create new surface features. Using this command, you individually select all the part surfaces you want copied.

Surf & Bnd

This is another method of selecting which part surfaces will be copied to create new surface features. Using this command, you select one of the part surfaces you want copied (called the seed surface) and the bounding surfaces.

NOTE: Surfaces and Surf & Bnd both provide access to options on the SURF GATHER sub-menu to further refine your selections.

PART ➡ FEATURE ➡ CREATE ➡ SURFACE ➡ TRIM

FEAT CLASS
Solid
Surface
Datum
Cosmetic
User Defined
Done/Return

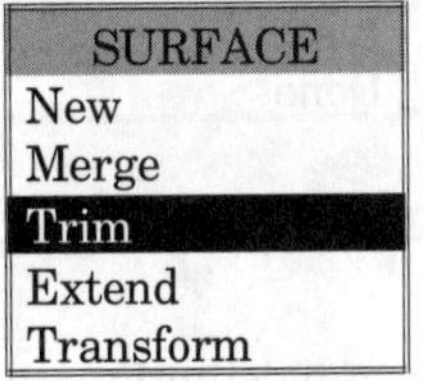

FORM
Extrude
Revolve
Sweep
Blend
Use Surfs
Silhouette
Advanced
Solid
Thin
Done
Quit

Silhouette

This command informs Pro/ENGINEER that you will be selecting another surface that will serve as the trimming plane for the surface to be trimmed.

Use Srfs

This command informs Pro/ENGINEER that you will be defining a viewing plane, and that the silhouette of the selected surface (as seen from the viewing plane) will serve as the trimming plane.

Feature - Group

PART ➡ FEATURE ➡ GROUP

PART
Feature
Modify
Regenerate
Relations
Family Tab
Interchange
Declare
Info
Interface
Set Up
Ref Dim
X-Section
Layer
Program

FEAT
Create
Pattern
Copy
Delete
Del Patern
Group
Supress
Resume
Reorder
Read Only
Redefine
Reroute
Mirror Geom
Insert Mode
Done

GROUP
Define
Make
Replace
Modify
Local Group
Ungroup
Pattern
Unpattern
List
Done/Return

Define

This command creates a group definition, then prompts you to identify which part features will be added to your group.

List

This command displays a list of all the groups which exist in the current working directory.

Local Group

Local groups vary from other group types in that you are not required to define placement parameters, and the group *can only be used in the current part.* This command is most often used to quickly associate multiple features so they can be patterned simultaneously.

Make

This command has the same functionality as the **FEAT** ➡ USER DEFINE command in part mode. Allows you to place a group (User Defined Feature) which has been previously defined and stored.

Modify

This command allows you to modify all the parameters of a group (User Defined Feature).

Pattern

This command allows you to pattern a group of features as if they were a single feature.

Replace

This command allows you to replace an occurance of a group with the contents of another group.

Ungroup

This command allows you to select a group occurance, placed using the **Make** command, and have Pro/ENGINEER remove the grouping parameter. The features that made up the group remain, but are no longer associated.

Unpattern

This command allows you to remove the effect of the **Pattern** command. Geometry stays, but dimensions are returned to a single reference rather than incremental.

PART ➡ FEATURE ➡ GROUP ➡ DEFINE

FEAT
Create
Pattern
Copy
Delete
Del Patern
Group
Supress
Resume
Reorder
Read Only
Redefine
Reroute
Mirror Geom
Insert Mode
Done

GROUP
Define
Make
Replace
Modify
Local Group
Ungroup
Pattern
Unpattern
List
Done/Return

DEFINE GROUP
Table
No Table

No Table

This command informs Pro/ENGINEER that you will be creating your group (User Defined Feature) manually. It will not have instances of the same generic.

Table

This command informs Pro/ENGINEER that you will be creating your group (User Defined Feature) using a family table.

Feature - Redefine

PART ➡ FEATURE ➡ REDEFINE

PART
Feature
Modify
Regenerate
Relations
Family Tab
Interchange
Declare
Info
Interface
Set Up
Ref Dim
X-Section
Layer
Program

FEAT
Create
Pattern
Copy
Delete
Del Patern
Group
Supress
Resume
Reorder
Read Only
Redefine
Reroute
Mirror Geom
Insert Mode
Done

REDEFINE
Attributes
Direction
Section
Flip
References
Boundaries
Scheme
Curves
Line Style
Corner Rounds
Placement
Pattern

Attributes

This command allows you to modify the *depth* and *intersection* attributes of a feature, such as (blind, thru next, thru all, etc.)

Boundaries

This command allows you to modify the extent of a surface feature created using **boundary** definitions (re-size a surface feature).

Corner Rounds

This command allows you to modify the *sphere radius* and *transitional distance* attributes of selected corner rounds.

Direction

This command allows you to modify the directional attributes of a feature, such as both sides, one side, etc.

Flip

This command allows you to modify the side of the section to which material is either added or subtracted.

Line Style

This command allows you to modify the line-style definitions for curves and cosmetic features.

References

This command allows you to modify the reference attributes of a feature, such as rounded edges. You will be prompted to specify which new edges should have rounds added, or from which existing edges the round definitions should be removed.

Scheme

This command allows you to modify the original sketched section's dimensioning scheme. Using this command, you will not be allowed to modify the actual sketched elements.

Section

This command allows you to modify the original sketched section used to create a feature. Be careful when using this command, as you may delete sketched elements currently used as parents of features added after the original feature. If you attempt to do, so Pro/ENGINEER will respond with a warning.

Feature - Reorder

PART ➡ FEATURE ➡ REORDER

PART
- Feature
- Modify
- Regenerate
- Relations
- Family Tab
- Interchange
- Declare
- Info
- Interface
- Set Up
- Ref Dim
- X-Section
- Layer
- Program

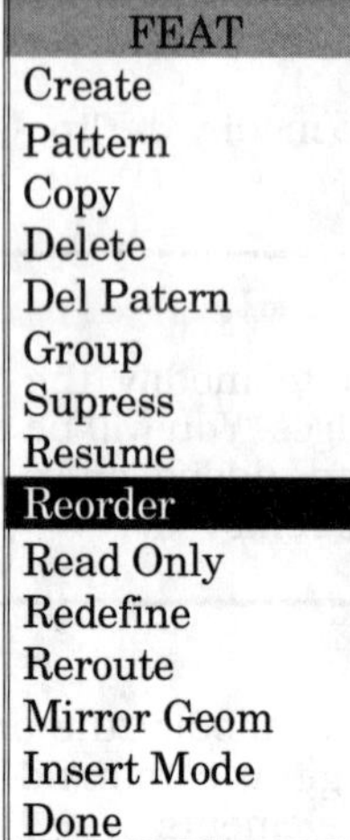

REORDER
- Earlier
- Later
- Select
- Feat Num
- Done/Return

Earlier

This command moves the desired feature forward in the feature list.

Feat Num

This command is used to select the feature you will be reordering, by specifying the **external ID** of the desired feature.

Later

This command moves the desired feature backward in the feature list.

Select

This command is used to select the feature you will be reordering from the screen.

Reference Dimension

PART ➥ REF DIM ➥ CREATE

PART
- Feature
- Modify
- Regenerate
- Relations
- Family Tab
- Interchange
- Declare
- Info
- Interface
- Set Up
- Ref Dim
- X-Section
- Layer
- Program

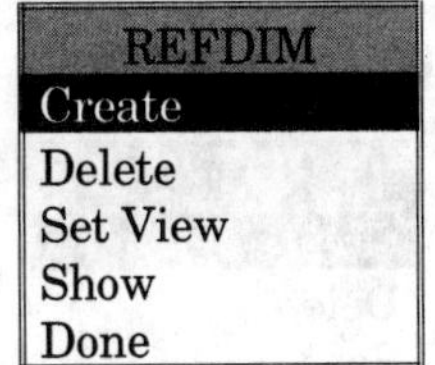

REF TYPE
- By Edge
- By Surface

By Edge

Creates a reference dimension between two selected edges. These dimensions are *view dependent* and must be associated with a named view. You may modify the view parameters of these dimensions using the **Set View** option.

By Surf

Creates a reference dimension between two selected surfaces. These dimensions are *view independent* and are automatically displayed in all views.

Create

This command is used to create a new reference dimension. You may do so by selecting a surface or an edge of a part.

Delete

This command is used to remove an existing reference dimension.

Set View

This command is used to modify the view parameters for reference dimensions created using the **By Edge** option.

Show

This command instructs Pro/ENGINEER to display *only* the reference dimensions of a selected feature.

PART ➡ REF DIM ➡ CREATE ➡ BY EDGE

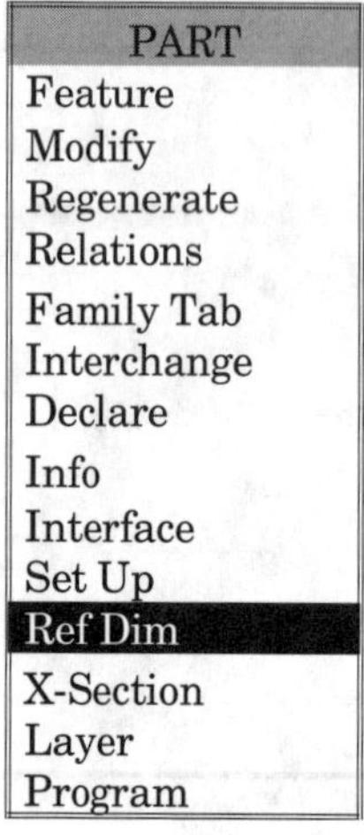

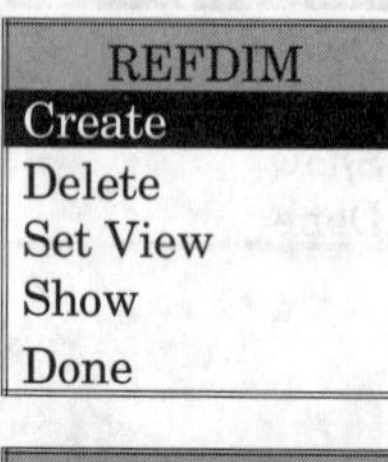

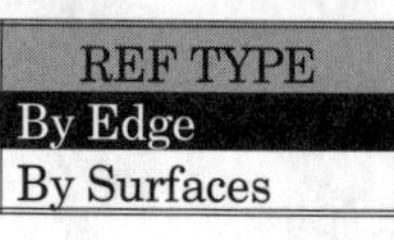

VIEW NAMES
Names
. . .

ATTACH TYPE
On Entity
Free Point
Midpoint
Intersect
Return

DIM ORIENT
Horizontal
Vertical
Slanted
Parallel
Normal

Attach Type

This sub-menu shows various placement options to further define the dimension's association to the selected edges.

Dim Orient

This sub-menu shows various placement options to further define the dimension's orientation and design intent.

Horizontal

This option instructs Pro/ENGINEER to create a reference dimension measuring the horizontal distance between the selected elements.

Intersect

The reference dimension will be attached to the closest intersection point of two selected elements.

Midpoint

The reference dimensions will be attached to the midpoint as it appears on the screen of the selected element.

Normal

This option instructs Pro/ENGINEER to create a reference dimension reflecting the distance normal to a selected element. You will be prompted to select an element to define the dimension's direction.

On Entity

The reference dimensions will be attached to the element at the selection point.

Parallel

This option instructs Pro/ENGINEER to create a reference dimension reflecting the distance parallel to a selected element. You will be prompted to select an element to define the dimension's direction.

Slanted

This option instructs Pro/ENGINEER to create a reference dimension measuring the shortest distance between two points. (This command is available only when the dimension is attached to points.)

Vertical

This option instructs Pro/ENGINEER to create a reference dimension measuring the vertical distance between the selected elements.

View Names

This sub-menu displays a name list of named views with which the reference dimension may be associated.

PART ➥ REF DIM ➥ SET VIEW

PART
Feature
Modify
Regenerate
Relations
Family Tab
Interchange
Declare
Info
Interface
Set Up
Ref Dim
X-Section
Layer
Program

REFDIM
Create
Delete
Set View
Show
Done

REF VIEW
Any View
Single View
Done

Any View

This command instructs Pro/ENGINEER to display the reference dimension in all views.

Single View

This command instructs Pro/ENGINEER to display the reference dimension *only* in the view in which it was created.

Relations

PART ➡ RELATIONS

PART
Feature
Modify
Regenerate
Relations
Family Tab
Interchange
Declare
Info
Interface
Set Up
Ref Dim
X-Section
Layer
Program

RELATIONS
Add
Edit Rel
Show Rel
Evaluate
Sort Rel
Add Param
Delete Param

Add

This command is used to add a relation in which one feature will be driven by the result of an equation. For example, d0 = d1+d2. When you use this command, you are prompted to enter the text for the equation. If you do not know the variables (d0,d1, etc.) for the features you wish to manipulate, use the **Modify** command to display the variables before running this command.

Add Param

This command is used to add *information* to your design without creating a relation. You may think of parameters as unique characteristics of the part (price, manufacturer, serial number, etc.).

Del Param

This command is used to remove any parameters you may have added to your model.

Edit Rel

This command is used to modify an existing relation. All the relations in your model will be displayed in a text editing window (using Pro/TABLE or the system editor, depending upon your configuration). Simply modify the values of the desired relations and save the file using the editor's command set.

Evaluate

This command is used to *test* the result of an equation without actually creating it. The result of your sample equation will be displayed in the message window. If the results are satisfactory, use the **Add** command to establish the relation.

Show Rel

This command is used to display all existing relations. Any parameters you have defined will also be displayed in the text window that will appear on your screen. You cannot edit relations using this command.

Sort Rel

This command is used to reorder the relations in your design, so that relations which are dependent upon the result of another are processed in the proper order during regeneration.

Setup

PART ➡ SETUP

PART
Feature
Modify
Regenerate
Relations
Family Tab
Interchange
Declare
Info
Interface
Set Up
Ref Dim
X-Section
Layer
Program

PART_SETUP
Material
Units
Mass Props
Dim Bound
Name
Done

Dim Bound

This command displays a sub-menu that allows you to modify selected feature dimensions, so the feature is displayed at its *nominal (ideal), upper (maximum),* or *lower (minimum)* tolerance values. You will use this command most often when performing a hypothetical interference detection.

Mass Props

This command allows you to create an ascii text file that contains mass properties information, and to assign those values to the part. All the properties in your model will be displayed in an information window (using Pro/TABLE or the system editor, depending upon your configuration).

Material

This command displays a sub-menu that allows you to create or edit material definitions that can be applied to the current part. Using this command, you can create a *library* of material definitions and perform "what-if" scenarios by quickly changing the material properties of the part.

Name

This command allows you to establish names for various items. You may create names for datum planes, coordinate systems, axes, curves, or features.

Units

This command allows you to specify the units of measure for both the *length* and the *mass* of your model.

PART ➥ SETUP ➥ MATERIAL

PART
Feature
Modify
Regenerate
Relations
Family Tab
Interchange
Declare
Info
Interface
Set Up
Ref Dim
X-Section
Layer
Program

PART_SETUP
Material
Units
Mass Props
Dim Bound
Name
Done

MATER_MGT
Define
Edit
Show
List
Assign
Unassign

Assign

This command is used to assign an existing material definition to the current part.

Define

This command is used to create a *new* material definition file, using the text-editing window provided. When the material definition is created and saved, the material properties are automatically assigned to the current part.

Edit

This command is used to modify an existing material definition file, using the text-editing window provided.

List

This command opens an information window and displays all the available material definitions.

Show

This command opens an information window and displays the properties of a specified material definition.

Unassign

This command is used to remove a material assignment from the current part.

Cross Section

PART ➥ X-SECTION

PART
Feature
Modify
Regenerate
Relations
Family Tab
Interchange
Declare
Info
Interface
Set Up
Ref Dim
X-Section
Layer
Program

XSEC ENTER
Create
Retrieve
Modify Xsec
Delete Xsec
Done / Return

Create

This command is used to create a new cross-section definition. Cross-sectional views are *slices* through your model (part or assembly), and are necessary for looking into the model for more detail.

Delete Xsec

This command is used to delete an existing cross-section definition. You define which section to delete by selecting its name from a displayed name list.

Modify Xsec

This command is used to modify an existing cross-section definition. There are many parameters controlling the appearance of a cross-section, and this command gives you access to the tools that change them.

PART ➡ X-SECTION ➡ CREATE

PART
Feature
Modify
Regenerate
Relations
Family Tab
Interchange
Declare
Info
Interface
Set Up
Ref Dim
X-Section
Layer
Program

XSEC ENTER
Create
Retrieve
Modify Xsec
Delete Xsec
Done / Return

XSEC CREATE
Planar
Offset
One Side
Both Sides
Done
Quit

XSEC OPTS
Model
Surf/Quit
Planar
Offset
Done
Quit

NOTE: The XSEC OPTS menu appears if your model has a surface.

Both Sides

Instructs Pro/ENGINEER that the sketched section will be projected in both directions from the sketching plane. The length of the projection will be determined by the specified **depth** option. Using this option, you may specify a unique depth option for each side. This option is only available with the **offset** method.

Offset

This option is used to create a cross-section along a defined path. Using this method, you create a two-dimensional sketch of the area you would like to see "opened-up," and the sketch will be extruded perpendicular to the sketching plane. Using this method, you can work your way through several key features of a part, instead of just going straight through the part at one depth.

One Side

Instructs Pro/ENGINEER that the sketched section will be projected in only one direction from the sketching plane. The length of the projection will be determined by the specified **depth** option. This option is only available with the **offset** method.

Planar

This option is used to create a cross-section along a datum plane. The plane can already exist or be created using the **Make Datum** command during the definition. This type of cross-section creates a single plane of information at a specified depth.

Surf / Quilt

This option is used to create a cross-section through a surface feature, to display its contour. This command also works with a selected or created datum plan that crosses through the surface feature.

PART ➥ X-SECTION ➥ MODIFY

PART
- Feature
- Modify
- Regenerate
- Relations
- Family Tab
- Interchange
- Declare
- Info
- Interface
- Set Up
- Ref Dim
- **X-Section**
- Layer
- Program

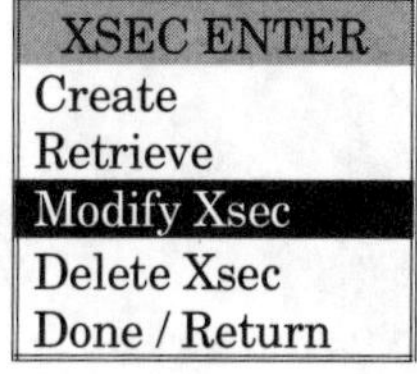

XSEC MODIFY
- Dim Values
- Redefine
- Hatching
- Name

Dim Values

This command allows you to change the various dimensions you previously assigned when sketching the section for an offset cross-section. After changing the values, you must regenerate the section. The dimensions do not have any effect on the part itself, they only change the shape of the cross-section's path.

Hatching

This command allows you to change the many parameters controlling the appearance of a cross-section's hatch pattern. Parameters such as spacing, angle, and style are easily modified using available submenus.

Name

All cross-section definitions must be named upon creation. This command allows you to change the name of an existing cross-section.

Redefine

This command goes beyond **Dim Values**, and allows you to re-create the entire sketch of an offset cross-section.

Section Five

ASSEMBLY

Assembly Commands

MODE ➡ ASSEMBLY

MAIN
Mode
Project
Dbms
Environment
Misc
Exit
Quit Window
ChangeWindow
View

MODE
Sketcher
Part
Sheet Metal
Composite
Assembly
Drawing
Manufacture
Mold
Layout
Format
Report
Markup
Diagram

ASSEMBLY
Component
Feature
Cabling
Modify
Regenerate
Relations
Family Tab
X-Section
Interchange
Declare
Info
Interface
Set Up
Ref Dim
Layer
Program

Component

This command displays a sub-menu allowing you to add or remove components from an assembly.

Family Table

This command provides access to a very powerful feature of Pro/ENGINEER called **Family Table** or "family of parts." Using this capability, you may create a *generic* part (assembly) that will be used as a template for creating similar (not identical) assemblies. The dimensions for the components and features of the resulting assemblies can be different from the original. The name **Family Table** comes from the fact that the resulting assemblies are *driven by* a table of dimensional values. (See Family Table in Section Three.)

Feature

This command displays a sub-menu allowing you to create assembly features (such as datum planes, coordinate systems, holes, etc.) which are extremely similar to part features, except that they belong to the assembly and not to any one component of the assembly.

Info

This command is used to display specific information about selected assembly components (such as mass properties, center of gravity, etc). In addition to the capabilities found in **Part** mode, you may generate a bill of materials using this command in **Assembly** mode.

Interchange

This command displays a sub-menu allowing you access to an extremely productive conceptual-design capability. Using the tools of this feature, you can place *preliminary* representations of an assembly component (simplified for speed). Once you are comfortable with the placement / design intent parameters, Pro/ENGINEER will automatically replace the conceptual component with the actual part or sub-assembly. This command may also be used to exchange one component for another.

Interface

This command displays a sub-menu allowing you to import and export data using various file formats.

Layer

Unlike other systems, which use layers in an overlay type of system to separate elements, **layers** are used in Pro/ENGINEER to group various elements together. Once the desired elements have been added to the **layer**, you may quickly display or turn off the display of all those grouped elements.

Modify

This command displays a sub-menu that allows you to modify the components of an assembly.

Ref Dim

Reference Dimensions are dimensions which do not have an impact on assembly data (they are not driving dimensions). They are placed on an assembly, part, or drawing for informational purposes only. They are *driven* by the assembly or part, and will be automatically adjusted during regeneration. (See Ref Dim in Section Three.)

Regenerate

To regenerate an assembly is to have Pro/ENGINEER evaluate the controlling dimensional schemes of the assembly components and modify them as necessary. See options for several methods of regenerating an assembly.

Relations

Relations are user defined equations that control the effect of component modifications. One feature of an assembly component may be related to another, using this capability. (See Relations in Section Three).

Setup

This command displays a sub-menu that allows you to control assembly properties such as units, names, mass property assignments, and clearance / interference values.

X-Section

This command allows you access to all the cross-section creation utilities.

Component

ASSEMBLY ➡ COMPONENT

ASSEMBLY
Component
Feature
Cabling
Modify
Regenerate
Relations
Family Tab
X-Section
Interchange
Declare
Info
Interface
Set Up
Ref Dim
Layer
Program

COMPONENT
Assemble
Disassemble
Represent
Package
Copy
Create
Merge
Cut Out
Delete
Del Pattern
Suppress
Resume
Pattern
Group
Reorder
Reroute
Insert Mode
Done

Assemble

This command is used to add a *component* to an assembly. Components are defined as **parts** or **sub-assemblies**. A sub-assembly is nothing more than a previously created assembly used as a component of another assembly. This command places the new component using parametric placement constraints, defining the component's relative location from existing components.

Copy

This command allows you to duplicate existing assembly components by creating a pattern. It requires you to select or create an assembly coordinate system and define the pattern's parameters.

Create

This command allows you to create a *part* while in **assembly** mode. Once saved with a part name, it can be accessed in **part** mode, used in other assemblies, and in general, will behave as any other Pro/ENGINEER part.

Cut Out

This command allows you to remove the overlapping material after two sets or parts have been placed in the assembly.

Del Pattern

This command is used to remove a pattern of existing assembly components, while maintaining the pattern's parent feature.

Delete

This command is used to remove an existing assembly component.

Disassemble

This command is used to remove a *component* from an assembly. Any components placed after the selected component will temporarily disappear until you decide the final fate of the selected component using the **Replace** options.

Group

This command is used to associate multiple components, so they may be manipulated as a single entity.

Insert Mode

This command is used to place new components at an earlier location in the component regeneration list.

Merge

This command allows you to add the material from one set of parts to another set, after both sets have been placed in the assembly.

Package

This command is used to place a component in a non-parametric manner, allowing for quick *what-if* modifications. Once the component is located in the desirable location, parametric references can be generated using the **Finalize** command.

Pattern

This command is used to make multiple duplications of an existing assembly component.

Reorder

This command is used to change the order that components are regenerated. You may not reorder a component or sub-assembly so that it is regenerated before any of it's parents.

Represent

This command works with the **Simplify** command to make complex objects appear with less detail in the display. Most often used to increase productivity when working with complex assemblies.

Reroute

This command is used to change the parent-child relationships established when the selected component or sub-assembly was placed. You are allowed to select new sketching, placement, and dimensioning reference features.

Resume

This command is used to **un-suppress** components that have been temporarily hidden using the **Suppress** command.

Suppress

This command is used to *temporarily* remove components from an assembly. Most commonly used to display the model in a simplified manner, to decrease the regeneration time requirements. Children of the **suppressed** component will not be displayed until the parent is resumed.

ASSEMBLY ➡ COMPONENT ➡ ASSEMBLE

ASSEMBLY
- Component
- Feature
- Cabling
- Modify
- Regenerate
- Relations
- Family Tab
- X-Section
- Interchange
- Declare
- Info
- Interface
- Set Up
- Ref Dim
- Layer
- Program

COMPONENT
- Assemble
- Disassemble
- Represent
- Package
- Copy
- Create
- Merge
- Cut Out
- Delete
- Del Pattern
- Suppress
- Resume
- Pattern
- Group
- Reorder
- Reroute
- Insert Mode
- Done

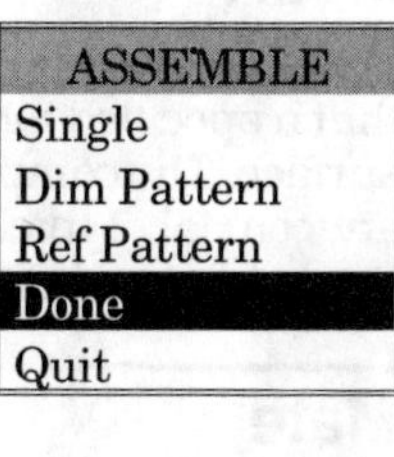

PLACE
- Mate
- Mate Offset
- Align
- Align Off
- Insert
- Orient
- Coord Sys
- Tangent
- Pnt On Srf
- Done
- Quit

Align

This command instructs Pro/ENGINEER to place the new component such that a specified planar surface is co-planar with a specified planar surface on an existing component. The new component will be oriented so the selected surface faces the same direction as the selected surface of the existing component.

Align Offset

This command instructs Pro/ENGINEER to place the new component such that a specified planar surface is offset *parallel* to a specified planar surface on an existing component. The new component will be oriented so the selected surface faces the same direction as the selected surface of the existing component.

Coord Sys

This command instructs Pro/ENGINEER to place the new component by aligning it's existing coordinate system with that of another coordinate system already in the assembly. The other coordinate system can be an existing **part** or **assembly** coordinate system.

Dim Pattern

Creates a pattern of the specified component using dimensional values to control the pattern.

Insert

This command instructs Pro/ENGINEER to place the components such that a specified *male* revolved surface is inserted into a specified *female* surface. The components will be oriented so the axes of the two surfaces are coaxial. Typically, this command option must be used in conjunction with another placement constraint.

Mate

This command instructs Pro/ENGINEER to place the new component such that a specified surface is coincident with a specified surface on an existing component. The new component will be oriented so the selected surface faces the selected surface of the existing component.

Mate Offset

This command instructs Pro/ENGINEER to place the new component such that a specified surface is offset *parallel* to a specified surface on an existing component. The new component will be oriented so the selected surface faces the selected surface of the existing component.

Orient

This command instructs Pro/ENGINEER to place the new component such that a specified planar surface is *parallel* to a specified planar surface on an existing component. The new component will be oriented so the selected surface faces the same direction as the selected surface of the existing component.

Pnt on Srf

This command instructs Pro/ENGINEER to place the new component such that a specified datum point is *in contact* with a specified surface on an existing component. Typically, this command option must be used in conjunction with another placement constraint.

Ref Pattern

Creates a pattern of the specified component that is dependent upon an existing *Dim Pattern* for its dimensional values.

Single

Places a single occurance of the component using the specified options.

Tangent

This command instructs Pro/ENGINEER to place the new component such that a specified surface is constrained at the tangencies with a specified surface on an existing component. Typically, this command option must be used in conjunction with another placement constraint.

ASSEMBLY ➥ COMPONENT ➥ COPY

ASSEMBLY
Component
Feature
Cabling
Modify
Regenerate
Relations
Family Tab
X-Section
Interchange
Declare
Info
Interface
Set Up
Ref Dim
Layer
Program

COMPONENT
Assemble
Disassemble
Represent
Package
Copy
Create
Merge
Cut Out
Delete
Del Pattern
Suppress
Resume
Pattern
Group
Reorder
Reroute
Insert Mode
Done

COPY
Translate
Rotate

Rotate

This command creates a pattern of duplicates from the selected assembly component by rotating the pattern about a specified coordinate system axis, using the values you specify.

Translate

This command creates a pattern of duplicates from the selected assembly component by translating the component in the direction of a coordinate system axis, using the values you specify.

ASSEMBLY ➥ COMPONENT ➥ CUT OUT and MERGE

ASSEMBLY	COMPONENT	MERGE OPT
Component	Assemble	Reference
Feature	Disassemble	Copy
Cabling	Represent	No Datums
Modify	Package	Copy Datums
Regenerate	Copy	Single
Relations	Create	Dim Pattern
Family Tab	Merge	Ref Pattern
X-Section	Cut Out	Done
Interchange	Delete	Quit
Declare	Del Pattern	
Info	Suppress	
Interface	Resume	
Set Up	Pattern	
Ref Dim	Group	
Layer	Reorder	
Program	Reroute	
	Insert Mode	
	Done	

Copy

This command instructs Pro/ENGINEER to copy all the features and relations associated with the *second* set of parts into the first.

Copy Datums

This command instructs Pro/ENGINEER to copy the datums of the *second* set of parts into the first.

Dim Pattern

This command instructs Pro/ENGINEER to create a pattern of **cut** or **merge** operations, using specified dimensional values to control the pattern. This command is only available when using the **Reference** option.

No Datums

This command instructs Pro/ENGINEER not to copy the datums of the *second* set of parts into the first. This command is only available when merging components using the **Reference** option.

Ref Pattern

This command instructs Pro/ENGINEER to create a pattern of **cut** or **merge** operations that are dependent upon an existing **Dim Pattern** for their dimensional values.

Reference

This command instructs Pro/ENGINEER to use the original *second* set of parts as referenced data. If the original component changes, the resulting addition or cut will be adjusted to match the change. However, if the original component is deleted or renamed you will experience regeneration problems and must re-establish the component, so use this command with careful attention.

Single

This command instructs Pro/ENGINEER to create a single occurance of the **cut** or **merge** operation.

ASSEMBLY ➡ COMPONENT ➡ DISASSEMBLE

ASSEMBLY
Component
Feature
Cabling
Modify
Regenerate
Relations
Family Tab
X-Section
Interchange
Declare
Info
Interface
Set Up
Ref Dim
Layer
Program

COMPONENT
Assemble
Disassemble
Represent
Package
Copy
Create
Merge
Cut Out
Delete
Del Pattern
Suppress
Resume
Pattern
Group
Reorder
Reroute
Insert Mode
Done

DISASSEMBLE
Last
Select
Done / Quit

Last

This command instructs Pro/ENGINEER to remove the last component placed in an assembly.

Select

This command instructs Pro/ENGINEER that you will be selecting which component to remove from an assembly.

ASSEMBLY ➡ COMPONENT ➡ DISASSEMBLE ➡ LAST or SELECT

ASSEMBLY
- Component
- Feature
- Cabling
- Modify
- Regenerate
- Relations
- Family Tab
- X-Section
- Interchange
- Declare
- Info
- Interface
- Set Up
- Ref Dim
- Layer
- Program

COMPONENT
- Assemble
- Disassemble
- Represent
- Package
- Copy
- Create
- Merge
- Cut Out
- Delete
- Del Pattern
- Suppress
- Resume
- Pattern
- Group
- Reorder
- Reroute
- Insert Mode
- Done

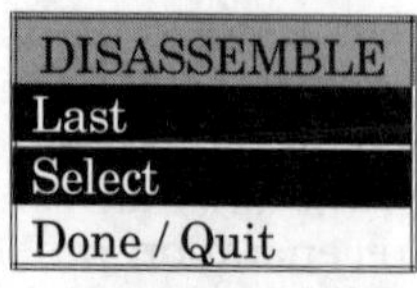

ASSY REPLACE
- Place Again
- Replace
- Remove
- Clip
- Place Back

Clip

This command instructs Pro/ENGINEER to keep the assembly in its current state and remove the selected component from the assembly, as well as any components placed after the selected item.

Place Again

This command instructs Pro/ENGINEER to restore the *dis-assembled* component using newly defined placement constraints. Components placed after the selected item are adjusted.

Place Back

This command instructs Pro/ENGINEER to restore the *dis-assembled* component, using its original placement constraints. Components placed after the selected item are returned to their original position also.

Replace

This command instructs Pro/ENGINEER to replace the selected component with a newly specified component.

Remove

This command instructs Pro/ENGINEER to remove the selected component from the assembly. It also removes any components placed after the selected item, so use with caution.

ASSEMBLY ➡ COMPONENT ➡ DISASSEMBLE ➡ PLACE AGAIN or REPLACE

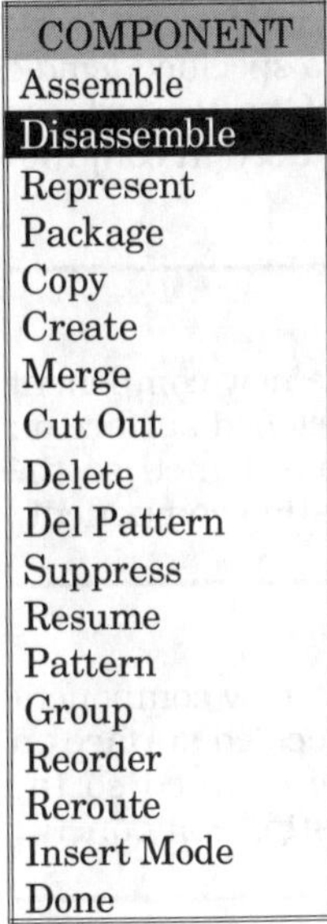

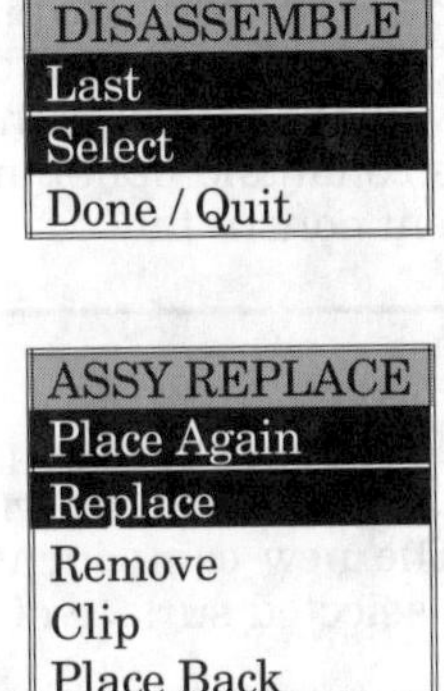

ASSEMBLE
Single
Dim Pattern
Ref Pattern
Done
Quit

PLACE
Mate
Mate Offset
Align
Align Off
Insert
Orient
Coord Sys
Tangent
Pnt On Srf
Done
Quit

Align

This command instructs Pro/ENGINEER to place the new component such that a specified planar surface is co-planar with a specified planar surface on an existing component. The new component will be oriented so the selected surface faces the same direction as the selected surface of the existing component.

Align Offset

This command instructs Pro/ENGINEER to place the new component such that a specified planar surface is offset *parallel* to a specified planar surface on an existing component. The new component will be oriented so the selected surface faces the same direction as the selected surface of the existing component.

Coord Sys

This command instructs Pro/ENGINEER to place the new component by aligning it's existing coordinate system with that of another coordinate system already in the assembly. The other coordinate system can be an existing **part** or **assembly** coordinate system.

Dim Pattern

Create a pattern of the specified component using dimensional values to control the pattern.

Insert

This command instructs Pro/ENGINEER to place the components such that a specified *male* revolved surface is inserted into a specified *female* surface. The components will be oriented so the axes of the two surfaces are coaxial. Typically, this command option must be used in conjunction with another placement constraint.

Mate

This command instructs Pro/ENGINEER to place the new component such that a specified surface is coincident with a specified surface on an existing component. The new component will be oriented so the selected surface faces the selected surface of the existing component.

Mate Offset

This command instructs Pro/ENGINEER to place the new component such that a specified surface is offset *parallel* to a specified surface on an existing component. The new component will be oriented so the selected surface faces the selected surface of the existing component.

Orient

This command instructs Pro/ENGINEER to place the new component such that a specified planar surface is *parallel* to a specified planar surface on an existing component. The new component will be oriented so the selected surface faces the same direction as the selected surface of the existing component.

Pnt on Srf

This command instructs Pro/ENGINEER to place the new component such that a specified datum point is *in contact* with a specified surface on an existing component. Typically, this command option must be used in conjunction with another placement constraint.

Ref Pattern

Creates a pattern of the specified component that is dependent upon an existing **Dim Pattern** for its dimensional values.

Single

Places a single occurance of the component, using the specified options.

Tangent

This command instructs Pro/ENGINEER to place the new component such that a specified surface is constrained at the tangencies with a specified surface on an existing component. Typically, this command option must be used in conjunction with another placement constraint.

ASSEMBLY ➡ COMPONENT ➡ PACKAGE

ASSEMBLY
Component
Feature
Cabling
Modify
Regenerate
Relations
Family Tab
X-Section
Interchange
Declare
Info
Interface
Set Up
Ref Dim
Layer
Program

COMPONENT
Assemble
Disassemble
Represent
Package
Copy
Create
Merge
Cut Out
Delete
Del Pattern
Suppress
Resume
Pattern
Group
Reorder
Reroute
Insert Mode
Done

PACKAGE
Place
Position
Finalize
Done / Quit

Finalize

This command is used to convert a *packaged* component to an assembled component using the **Assemble** options.

Place

This command is used to place a component in an arbitrary location, determined by the system.

Position

This command is used to relocate a *packaged* component by moving or rotating it.

ASSEMBLY ➡ COMPONENT ➡ PACKAGE ➡ FINALIZE

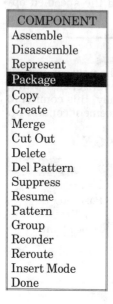

COMPONENT
Assemble
Disassemble
Represent
Package
Copy
Create
Merge
Cut Out
Delete
Del Pattern
Suppress
Resume
Pattern
Group
Reorder
Reroute
Insert Mode
Done

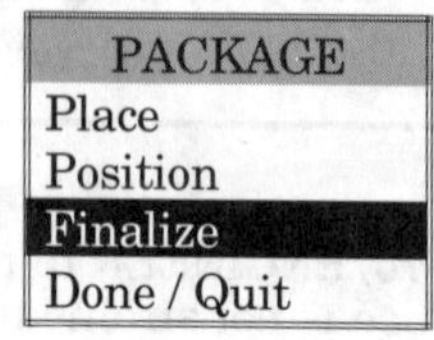

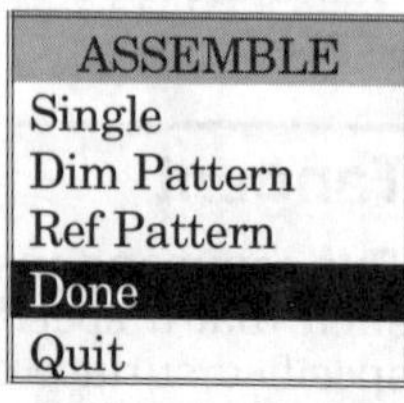

PLACE
Mate
Mate Offset
Align
Align Offset
Insert
Orient
Coord Sys
Tangent
Pnt On Srf
Done
Quit

Align

This command instructs Pro/ENGINEER to place the new component such that a specified planar surface is co-planar with a specified planar surface on an existing component. The new component will be oriented so the selected surface faces the same direction as the selected surface of the existing component.

Align Offset

This command instructs Pro/ENGINEER to place the new component such that a specified planar surface is offset *parallel* to a specified planar surface on an existing component. The new component will be oriented so the selected surface faces the same direction as the selected surface of the existing component.

Coord Sys

This command instructs Pro/ENGINEER to place the new component by aligning it's existing coordinate system with that of another coordinate system already in the assembly. The other coordinate system can be an existing **part** or **assembly** coordinate system.

Dim Pattern

Create a pattern of the specified component, using dimensional values to control the pattern.

Insert

This command instructs Pro/ENGINEER to place the components such that a specified *male* revolved surface is inserted into a specified *female* surface. The components will be oriented so the axes of the two surfaces are coaxial. Typically, this command option must be used in conjunction with another placement constraint.

Mate

This command instructs Pro/ENGINEER to place the new component such that a specified surface is coincident with a specified surface on an existing component. The new component will be oriented so the selected surface faces the selected surface of the existing component.

Mate Offset

This command instructs Pro/ENGINEER to place the new component such that a specified surface is offset *parallel* to a specified surface on an existing component. The new component will be oriented so the selected surface faces the selected surface of the existing component.

Orient

This command instructs Pro/ENGINEER to place the new component such that a specified planar surface is *parallel* to a specified planar surface on an existing component. The new component will be oriented so the selected surface faces the same direction as the selected surface of the existing component.

Pnt on Srf

This command instructs Pro/ENGINEER to place the new component such that a specified datum point is *in contact* with a specified surface on an existing component. Typically this command option must be used in conjunction with another placement constraint.

Ref Pattern

Creates a pattern of the specified component that is dependent upon an existing **Dim Pattern** for its dimensional values.

Single

Places a single occurance of the component using the specified options.

Tangent

This command instructs Pro/ENGINEER to place the new component such that a specified surface is constrained at the tangencies with a specified surface on an existing component. Typically, this command option must be used in conjunction with another placement constraint.

ASSEMBLY ➡ COMPONENT ➡ PACKAGE ➡ POSITION

COMPONENT
Assemble
Disassemble
Represent
Package
Copy
Create
Merge
Cut Out
Delete
Del Pattern
Suppress
Resume
Pattern
Group
Reorder
Reroute
Insert Mode
Done

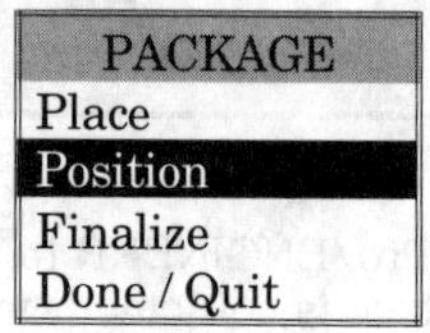

PLACE COMP
Move
Rotate

Move

This command displays a sub-menu that allows you to move a *packaged* component parallel or normal to a plane or edge.

Rotate

This command displays a sub-menu allowing you to rotate a *packaged* component about a selected point, edge or axis.

ASSEMBLY ➥ COMPONENT ➥ PACKAGE ➥ POSITION ➥ MOVE

COMPONENT
Assemble
Disassemble
Represent
Package
Copy
Create
Merge
Cut Out
Delete
Del Pattern
Suppress
Resume
Pattern
Group
Reorder
Reroute
Insert Mode
Done

PACKAGE
Place
Position
Finalize
Done / Quit

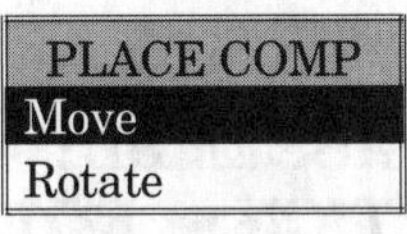

MOVE COMP
Along View
Along Edge
Along Axis
Along Face
Norm Plane
Move Until
Done

Along Axis

This command translates the selected component(s) parallel to a selected axis.

Along Edge

This command translates the selected component(s) parallel to a selected edge.

Along Face

This command translates the selected component(s) parallel to a selected plane.

Along View

This command translates the selected component(s) in the plane of the view.

Move Until

This command translates the selected component(s) until the selected planes become co-planar, or the selected axes become co-axial.

Norm Plane

This command translates the selected component(s) normal to a selected plane.

ASSEMBLY ➥ COMPONENT ➥ PACKAGE ➥ POSITION ➥ ROTATE

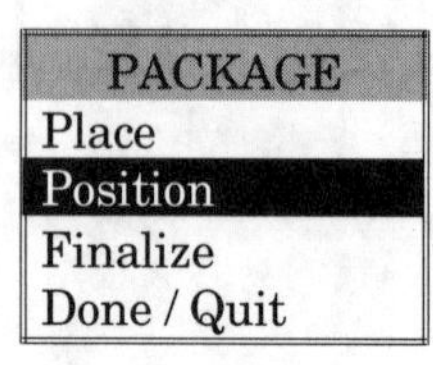

ROTATE COMP
View Point
About Edge
About Axis
Rot Until
Done

About Axis

This command rotates the selected component(s) about a selected axis.

About Edge

This command rotates the selected component(s) about a selected edge.

Rot Until

This command rotates the selected component(s) until the selected planes become co-planar, or the selected axes become co-axial.

View Point

This command rotates the selected component(s) about a selected point on the viewing plane.

Feature

ASSEMBLY ➡ FEATURE

ASSEMBLY
Component
Feature
Cabling
Modify
Regenerate
Relations
Family Tab
X-Section
Interchange
Declare
Info
Interface
Set Up
Ref Dim
Layer
Program

ASSY FEAT
Create
Intersect
Pattern
Group
Delete
Del Pattern
Suppress
Resume
Redefind
Reorder
Reroute
Done / Return

Create

This command allows you to create *features* while in **Assembly** mode. Unlike part features, these features (holes, cuts, etc.) may effect several components placed in the assembly. These features will only be available in the assembly mode, and do not physically modify the original components.

Del Pattern

This command is used to remove a pattern of existing assembly features, while maintaining the pattern's parent feature.

Delete

This command is used to remove an existing assembly feature.

Group

This command is used to associate multiple assembly features, so they may be manipulated as a single entity.

Intersect

This command is used to identify which assembly components will or will not be affected by the assembly feature.

Pattern

This command is used to make multiple duplications of an existing assembly feature.

Redefine

This command displays a sub-menu of options that allow you to modify *how* an assembly feature was originally created.

Reorder

This command is used to change the order in which assembly features are regenerated. You may not reorder an assembly feature so that it is regenerated before any of its parents.

Reroute

This command is used to change the parent-child relationships established when the selected assembly feature was created. You are allowed to select new sketching, placement, and dimensioning reference features.

Resume

This command is used to **un-suppress** assembly features that have been temporarily hidden using the **Suppress** command.

Suppress

This command is used to *temporarily* remove assembly features from an assembly. Most commonly used to display the model in a simplified manner, to decrease the regeneration time requirements. Children of the **suppressed** component will not be displayed until the parent is resumed.

Modify

ASSEMBLY ➥ MODIFY

ASSEMBLY
Component
Feature
Cabling
Modify
Regenerate
Relations
Family Tab
X-Section
Interchange
Declare
Info
Interface
Set Up
Ref Dim
Layer
Program

ASSEM MOD
Mod Part
Mod Assem
Mod Dim
Mod Expld
Edit Expld
Done / Return

Edit Expld

This command allows you to further define how the various components of an assembly will be separated when the **Exploded View** option is chosen. This offers functionality not found in the **Mod Expld** option.

Mod Expld

This command controls the default distances that the various components will be separated by, when the **Exploded View** option is chosen.

Mod Part

This command is used to create, delete, modify, or suppress any part features while in assembly mode.

ASSEMBLY ➡ MODIFY ➡ MOD PART

ASSEMBLY
Component
Feature
Cabling
Modify
Regenerate
Relations
Family Tab
X-Section
Interchange
Declare
Info
Interface
Set Up
Ref Dim
Layer
Program

ASSEM MOD
Mod Part
Mod Assem
Mod Dim
Mod Expld
Edit Expld
Done / Return

MODIFY PART
Feature
Modify Dim
Regenerate
Layer
Done

Feature

This command is used to create, delete, modify, etc. features of a part, but not assembly features.

Layer

This command is used to add part features to layer definitions so the contents of that layer may be quickly hidden and recalled as necessary.

Mod Dim

This command is used to modify the dimensional values of any part while in assembly mode.

Regenerate

Any parts modified using the **Mod Dim** option must be regenerated using this command before the changes will be incorporated.

ASSEMBLY ➡ MODIFY ➡ EDIT EXPLD

ASSEMBLY
Component
Feature
Cabling
Modify
Regenerate
Relations
Family Tab
X-Section
Interchange
Declare
Info
Interface
Set Up
Ref Dim
Layer
Program

ASSEM MOD
Mod Part
Mod Assem
Mod Dim
Mod Expld
Edit Expld
Done / Return

EDIT EXPLD
Add
Remove

Add

This command instructs Pro/ENGINEER to add specific exploded view location instructions on a selected component. You may move a component normal to a selected plane or tangent to an axis or edge.

Remove

This command instructs Pro/ENGINEER to delete specific exploded view location instructions on a selected component. After this command is executed, the components will be located using the default values assigned with the **Mod Expld** option.

Regenerate

ASSEMBLY ➡ REGENERATE

ASSEMBLY
Component
Feature
Cabling
Modify
Regenerate
Relations
Family Tab
X-Section
Interchange
Declare
Info
Interface
Set Up
Ref Dim
Layer
Program

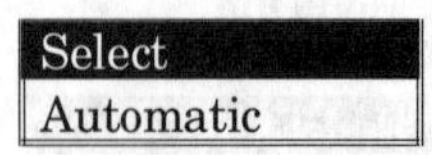

SELECT PART
Pick Part
Layer
Upd Int Prts
Don't Update
Simple
Recursive
Done
Quit

Automatic

This command instructs Pro/ENGINEER to update *every* assembly component that has not been regenerated since the last modification. (No selection is required).

Don't Update

This command comes into play when you have **assembly features** in your model. It instructs Pro/ENGINEER *not* to regenerate any components intersected by the assembly feature.

Layer

This command instructs Pro/ENGINEER that you will be selecting a layer name, and want all the assembly components associated with that layer to be regenerated.

Pick Part

This command instructs Pro/ENGINEER that you will be selecting components (to be regenerated) from the screen, or from the name list.

Recursive

This command comes into play when you have assembly components with **external references** in your model. It instructs Pro/ENGINEER to regenerate the external references first, and then use the updated values to regenerate the components.

Select

This command instructs Pro/ENGINEER to update *only* those assembly components that you specifically identify.

Simple

This command comes into play when you have assembly components with **external references** in your model. It instructs Pro/ENGINEER to regenerate those components using the last known values for the external references.

Upd Int Prts

This command comes into play when you have **assembly features** in your model. It instructs Pro/ENGINEER to regenerate *all* components intersected by the assembly feature.

Relations

ASSEMBLY ➡ RELATIONS

ASSEMBLY
Component
Feature
Cabling
Modify
Regenerate
Relations
Family Tab
X-Section
Interchange
Declare
Info
Interface
Set Up
Ref Dim
Layer
Program

MODEL REL
Assem Rel
Part Rel
Sket Rel
Done

Assem Rel

This command instructs Pro/ENGINEER that you will be creating or modifying relations that govern how the components of an assembly will behave.

Part Rel

This command instructs Pro/ENGINEER that you will be creating or modifying relations that govern how the features of a single part will behave.

Sket Rel

This command instructs Pro/ENGINEER that you will be creating or modifying relations that govern how the elements of an assembly or part feature will behave.

ASSEMBLY ➡ RELATIONS ➡ ASSEM REL

ASSEMBLY
Component
Feature
Cabling
Modify
Regenerate
Relations
Family Tab
X-Section
Interchange
Declare
Info
Interface
Set Up
Ref Dim
Layer
Program

MODEL REL
Assem Rel
Part Rel
Sket Rel
Done

RELATIONS
Add
Edit Rel
Show Rel
Sort Rel
Add Param
Del Param
User Prog
Where Used

Add

This command is used to add a relation in which one component will be driven by the result of an equation. For example, d0 = d1+d2. When you use this command you are prompted to enter the text for the equation. If you do not know the variables (d0,d1, etc.) for the features you wish to manipulate, use the **Modify** command to display the variables, before running this command.

Add Param

This command is used to add *information* to your design without creating a relation. You may think of parameters as unique characteristics of the part (price, manufacturer, serial number, etc.).

Del Param

This command is used to remove any parameters you may have added to your model.

Edit Rel

This command is used to modify an existing relation. All the relations in your model will be displayed in a text editing window (using Pro/TABLE or the system editor, depending upon your configuration). Simply modify the values of the desired relations and save the file using the editor's command set.

Evaluate

This command is used to *test* the result of an equation without actually creating it. The result of your sample equation will be displayed in the message window. If the results are satisfactory, use the **Add** command to establish the relation.

Show Rel

This command is used to display all existing relations. Any parameters you have defined will also be displayed in the text window which will appear on your screen. You cannot edit relations using this command.

Sort Rel

This command is used to reorder the relations in your design so that relations that are dependent upon the result of another, are processed in the proper order during regeneration.

Setup

ASSEMBLY ➡ SETUP

ASSEMBLY
Component
Feature
Cabling
Modify
Regenerate
Relations
Family Tab
X-Section
Interchange
Declare
Info
Interface
Set Up
Ref Dim
Layer
Program

ASSEM SETUP
Mass Props
Units
Dim Bound
Done

Dim Bound

This command displays a sub-menu that allows you to modify selected assembly component dimensions so the component is displayed at its *nominal (ideal), upper (maximum)* or *lower (minimum)* tolerance value. You will use this command most often when performing a hypothetical interference detection.

Mass Props

This command allows you to create an ascii text file that contains mass properties information, and to assign those values to the assembly components. All the properties in your model will be displayed in a text editing window (using Pro/TABLE or the system editor, depending upon your configuration).

Name

This command allows you to establish names for various assembly items. You may create names for assembly components, coordinate systems, axes, curves, or features.

Units

This command allows you to specify the units of measure for both the *length* and the *mass* of your model.

ASSEMBLY ➡ SETUP ➡ DIM BOUND

ASSEMBLY
Component
Feature
Cabling
Modify
Regenerate
Relations
Family Tab
X-Section
Interchange
Declare
Info
Interface
Set Up
Ref Dim
Layer
Program

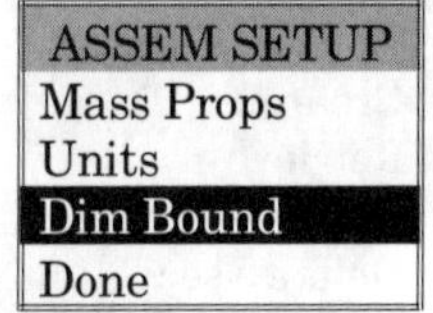

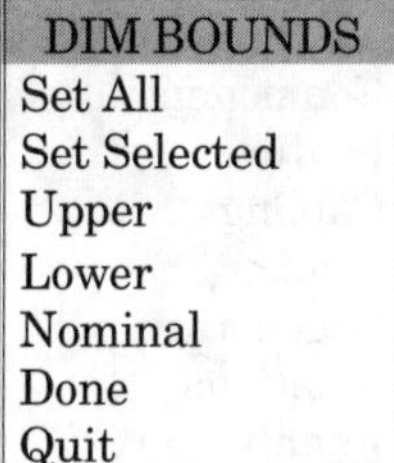

Lower

This command instructs Pro/ENGINEER to display the selected component(s) or sub-assemblies using the *minimum* tolerance value.

Nominal

This command instructs Pro/ENGINEER to display the selected component(s) or sub-assemblies using the *ideal* tolerance value.

Set All

This command instructs Pro/ENGINEER to set *all* dimensions to the specified boundary condition (nominal, upper, or lower).

Set selected

This command instructs Pro/ENGINEER to set only the dimensions you choose to the specified boundary condition (nominal, upper, or lower).

Upper

This command instructs Pro/ENGINEER to display the selected component(s) or sub-assemblies using the *maximum* tolerance value.

Cross Section

ASSEMBLY ➥ X-SECTION — Options

ASSEMBLY
Component
Feature
Cabling
Modify
Regenerate
Relations
Family Tab
X-Section
Interchange
Declare
Info
Interface
Set Up
Ref Dim
Layer
Program

XSEC ENTER
Create
Retrieve
Modiy Xsec
Delete Xsec
Done / Return

Create

This command is used to create a new cross-section definition. Cross-sectional views are *slices* through your model (part or assembly), and are necessary for looking into the model for more detail.

Delete Xsec

This command is used to delete an existing cross-section definition. You define which section to delete by selecting its name from a displayed name list.

Modify Xsec

This command is used to modify an existing cross-section definition. There are many parameters controlling the appearance of a cross-section, and this command gives you access to the tools with which to change them.

ASSEMBLY ➡ X-SECTION ➡ CREATE

ASSEMBLY	XSEC ENTER	XSEC CREATE
Component	**Create**	Planar
Feature	Retrieve	Offset
Cabling	Modiy Xsec	One Side
Modify	Delete Xsec	Both Sides
Regenerate	Done / Return	Done
Relations		Quit
Family Tab		
X-Section		
Interchange		
Declare		
Info		
Interface		
Set Up		
Ref Dim		
Layer		
Program		

XSECT OPTS
Model
Surf / Quilt
Planar
Offset
Done
Quit

NOTE: The XSEC OPTS menu only appears if there is a surface in your model.

Both Sides

Instructs Pro/ENGINEER that the sketched section will be projected in both directions from the sketching plane. The length of the projection will be determined by the specified *depth option.* Using this option, you may specify a unique depth option for each side. This option is only available with the **offset** method.

Offset

This option is used to create a cross-section along a defined path. Using this method, you create a two-dimensional sketch of the area you would like to see "opened-up" and the sketch will be extruded perpendicular to the sketching plane. Using this method, you can work your way through several key features of an assembly component, instead of going straight through the part at one depth.

One Side

Instructs Pro/ENGINEER that the sketched section will be projected in only one direction from the sketching plane. The length of the projection will be determined by the specified *depth option.* This option is only available with the **offset** method.

Planar

This option is used to create a cross-section along a datum plane. The plane can already exist or can be created using the **Make Datum** command during the definition. This type of cross-section creates a single plane of information at a specified depth.

Surf / Quilt

This option is used to create a cross-section through a surface feature, to display its contour. This command also works with a selected or created datum plan that crosses through the surface feature.

ASSEMBLY ➡ X-SECTION ➡ MODIFY

ASSEMBLY
Component
Feature
Cabling
Modify
Regenerate
Relations
Family Tab
X-Section
Interchange
Declare
Info
Interface
Set Up
Ref Dim
Layer
Program

XSEC ENTER
Create
Retrieve
Modiy Xsec
Delete Xsec
Done / Return

XSEC MODIFY
Dim Values
Redefine
Hatching
Name

Dim Values

This command allows you to change the various dimensions you previously assigned when sketching the section for an offset cross-section. After changing the values, you must regenerate the section. The dimensions do not have any effect on the part itself, they only change the shape of the cross-section's path.

Hatching

This command allows you to change the many parameters controlling the appearance of a cross-section's hatch pattern. Parameters such as spacing, angle, and style are easily modified using available sub-menus.

Name

All cross-section definitions must be named upon creation. This command allows you to change the name of an existing cross-section.

Redefine

This command goes beyond **Dim Values**, and allows you to re-create the entire sketch of an offset cross-section.

[illegible] do not have any effect on the part itself they only change the shape of the cross-section [illegible].

[illegible]

[illegible]

[illegible] allowed you to change the name of an existing [illegible]

[illegible]

[illegible] and allows you [illegible]

DRAWING

Drawing Commands

MODE ➥ DRAWING ➥ RETRIEVE

MAIN
Mode
Project
Dbms
Environment
Misc
Exit
Quit Window
ChangeWindow
View

MODE
Sketcher
Part
Sheet Metal
Composite
Assembly
Drawing
Manufacture
Mold
Layout
Format
Report
Markup
Diagram

ENTER DRAWING
Create
Retrieve
List
Import
Search / Retr

DRAWING
Views
Sheets
Modify
Regenerate
Switch Dim
Relations
Detail
Interface
Dwg Format
Table
Layer
Set Up
User Attrbt
Symbol
Set Model
Info
Represent

Create

This command is used to create a *new* drawing file in the current directory.

Clean Dims

This command is used to modify the placement of existing dimensions. Pro/ENGINEER will automatically rearrange the dimension elements to present a more professional image.

Detail

This command is used add dimensions, symbols, notes, and geometric tolerance symbols to your drawing file or format.

Dwg Format

This command is used to recall a pre-created drawing standard at any time. The *format* file can contain parameters for borders, text heights, fonts, title block attributes, tables and much more. Drawing formats may be parametric (created in **Sketcher** mode) or non-parametric (created in **Format** mode).

Info

This command is used to display specific information about selected items (such as mass properties, center of gravity, list of file names, draft elements by layer or type, etc.).

Interface

This command takes you to the Interface menu, for file import / export capabilities.

Layer

Unlike other systems, which use layers in an overlay type of system to separate elements, **layers** are used in Pro/ENGINEER to group various elements together. Once the desired elements have been added to the **layer**, you may quickly display or turn off the display of all those grouped elements.

Modify

This command allows for the manipulation of various drawing element types. The function of this command varies by the element type selected.

Regenerate

This command is used to update the model associated with a drawing file, any associative dimensions placed in your drawing file, or both.

Represent

This command works with the **Simplify** and **Restore** commands to modify the appearance of a part's features or an assembly's components.

Retrieve

This command is used to recall an existing drawing file.

Set Model

When working with a drawing file that has more than one model attached, this command is used to specify which model is the *active* model.

Set Up

Drawing files in Pro/ENGINEER have an external setup file which contains information governing the drawing's numerous parameters (fonts, types of arrow heads, etc.). The parameters are too numerous to mention in this text, so please refer to Appendix A of Parametric Technology Inc.'s ***Drawing Users Guide***.

Sheets

This command is used to create a *multiple-sheet* drawing, and allows you to move information from one sheet to another.

Switch Dim

This command is used to change the display of a selected dimension. If numerical values are shown, this command will modify the display to the symbolic name of the dimension (d0,d1) and vice versa.

Symbol

A symbol is a collection of draft entities and/or notes, that have been grouped together to act as a single element. You may use the options for this command to create and place *simple* or *generic* symbols. Simple symbols are simply duplicates of the original. Generic symbols are very similar to **User Defined Features** and allow you to place variations of the generic symbol template.

Table

This command displays a sub-menu allowing you to create a *spread-sheet-like* chart, consisting of horizontal rows and vertical columns (called **cells**), which contain information. The information can either be in the form of simple text or *automatically updating* information such as dimension symbols and drawing labels.

User Attrbt

This command is used to store information about a drawing file without creating a note. You may store information in the form of *text, floating point,* or *integer* data types. Each attribute must have a name-value combination.

Views

Views are the means by which you attach your model(s) to a drawing file. This command displays a sub-menu that allows you to add or modify a view of your attached model. The first time you use this command you must specify the model for which you will be creating a view.

Create

DRAWING ➡ CREATE

MODE
Sketcher
Part
Sheet Metal
Composite
Assembly
Drawing
Manufacture
Mold
Layout
Format
Report
Markup
Diagram

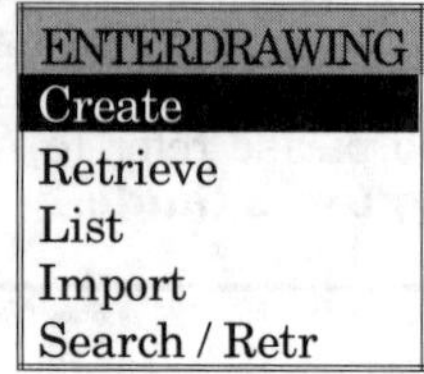

GET FORMAT
Set Dwg Size
Retr Format

Retr Format

This command is used to recall a pre-created drawing standard at the time of drawing creation. The **format** file can contain parameters for borders, text heights, fonts, title block attributes, tables, and much more. The format's parameters will be applied to the new file.

Set Dwg Size

This command displays a sub-menu of standard drawing sizes (A0-F) and orientations (landscape / portrait). If your required drawing size is not available, you may use the **Variable** option to create a drawing file of any size.

DRAWING ➥ CREATE ➥ SET DWG SIZE

MODE
- Sketcher
- Part
- Sheet Metal
- Composite
- Assembly
- **Drawing**
- Manufacture
- Mold
- Layout
- Format
- Report
- Markup
- Diagram

ENTERDRAWING
- **Create**
- Retrieve
- List
- Import
- Search / Retr

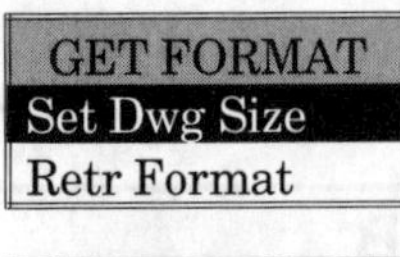

DWG SIZE TYP
- Portrait
- Landscape
- Variable

DWG SIZE
- A0
- A1
- A2
- A3
- A4
- F
- E
- D
- C
- B
- A

A0

This command instructs Pro/ENGINEER that you desire a 841 by 1189 mm drawing sheet.

A1

This command instructs Pro/ENGINEER that you desire a 594 by 841 mm drawing sheet.

A2

This command instructs Pro/ENGINEER that you desire a 420 by 594 mm drawing sheet.

A3

This command instructs Pro/ENGINEER that you desire a 297 by 420cmm drawing sheet.

A4

This command instructs Pro/ENGINEER that you desire a 210 by 297 mm drawing sheet.

A

This command instructs Pro/ENGINEER that you desire a 8.5 inch by 11 inch drawing sheet.

B

This command instructs Pro/ENGINEER that you desire a 11 inch by 17 inch drawing sheet.

C

This command instructs Pro/ENGINEER that you desire a 17 inch by 22 inch drawing sheet.

D

This command instructs Pro/ENGINEER that you desire a 22 inch by 34 inch drawing sheet.

E

This command instructs Pro/ENGINEER that you desire a 34 inch by 44 inch drawing sheet.

F

This command instructs Pro/ENGINEER that you desire a 28 inch by 40 inch drawing sheet.

Landscape

This command instructs Pro/ENGINEER that the drawing sheet will be oriented in such a manner that the larger of the sheet's two dimensions will define the horizontal value.

Portrait

This command instructs Pro/ENGINEER that the drawing sheet will be oriented in such a manner that the larger of the sheet's two dimensions will define the vertical value.

Variable

This command instructs Pro/ENGINEER that the horizontal and vertical values will be independently defined by you.

Detail

DRAWING ➡ RETRIEVE ➡ DETAIL

ENTERDRAWING
Create
Retrieve
List
Import
Search / Retr

DRAWING
Views
Sheets
Modify
Regenerate
Switch Dim
Relations
Detail
Interface
Dwg Format
Table
Layer
Set Up
User Attrbt
Symbol
Set Model
Info
Represent
Clean Dims

DETAIL
Show
Erase
Create
Delete
Move
Move Text
Move Attach
Break
Clip
Switch View
Flip Arrows
Make Job
Sketch
Tools
Modify
Done / Return

Break

This command is used to create dimension-witness lines and note-leader lines. You cannot combine **jogs** and **breaks** on the same line.

Delete

This command is used to permanently remove a selected item.

Erase

This command is used to temporarily blank a selected item, without actually removing it from the drawing file. Items that have been erased may be recalled using the **Show** command.

Flip Arrows

This command is used to reverse the direction of selected dimension arrows.

Make Jog

This command is used to create a connected offset (jog) in dimension-witness lines and note-leader lines. You cannot combine **jogs** and **breaks** on the same line.

Modify

This command will display a sub-menu that allows you to change entities of various types. Choose the type of element you wish to change to display the options for that element type.

Move Attach

This command is used to relocate the attachment point of a leader line.

Move Text

This command is used to lengthen or shorten the length of a leader line's elbow by moving the text in a horizontal manner. You may also use this command to change which side of the leader line the elbow is on.

Show

This command is used to display information about items in the drawing (dimensions, notes, balloons, symbols, etc.). The information may not be currently displayed, or may have been previously erased.

Sketch

This command is similar to the **Sketch** command in **Sketcher** mode. It is used to arbitrarily place draft geometry in a drawing file.

Switch View

This command is used to move selected items from one view to another.

Detail - Break

DRAWING ➥ RETRIEVE ➥ DETAIL ➥ BREAK

DRAWING
Views
Sheets
Modify
Regenerate
Switch Dim
Relations
Detail
Interface
Dwg Format
Table
Layer
Set Up
User Attrbt
Symbol
Set Model
Info
Represent
Clean Dims

DETAIL
Show
Erase
Create
Delete
Move
Move Text
Move Attach
Break
Clip
Switch View
Flip Arrows
Make Job
Sketch
Tools
Modify
Done / Return

BREAK
Add
Delete

BREAK TYPE
Dimension
Simple

Dimension

This command is used to create a gap at the intersection of two witness lines. The gap length will be determined by the settings in the drawing setup file.

Simple

This command is used to create a gap with an arbitrary length at any point along a witness or leader line.

Detail - Create

DRAWING ➥RETRIEVE ➥DETAIL ➥ CREATE & SHOW

DRAWING	DETAIL	DETAIL ITEM
Views	**Show**	Dimension
Sheets	Erase	Ref Dim
Modify	**Create**	Note
Regenerate	Delete	Geom Tol
Switch Dim	Move	Surf Finish
Relations	Move Text	Datum
Detail	Move Attach	X-section
Interface	Break	Balloon
Dwg Format	Clip	Axis
Table	Switch View	Symbol
Layer	Flip Arrows	
Set Up	Make Job	
User Attrbt	Sketch	
Symbol	Tools	
Set Model	Modify	
Info	Done / Return	
Represent		
Clean Dims		

Axis

This command is used to display a **draft axis** which is a draft entity contained in the drawing file. To create a draft axis, choose the command and select a cylindrical or conical surface.

Balloon

This command is used to create a note that is enclosed by a circle. The circles may be attached to a model edge of "free floating" components. The notes may consist of a single character or many. The size of the circle will be determined by the size requirements for the longest note. You may create balloon notes that are stacked in a horizontal or vertical alignment using the available options.

Datum

This command is used to create a datum plane within a drawing file.

Dimension

This command is used to create draft dimensions using the various options available.

Geom Tol

This command is used to place symbols and notes that adhere to the geometric dimensioning and tolerance standard.

Note

This command is used to create annotation in a drawing file using the various options available.

Ref Dimension

This command is used to create a special type of dimension, called a *reference* dimension, in the drawing file. Reference dimensions are used to display dimensional values that would not normally be displayed with the part dimensions.

Surf Finish

This command is used to place surface finish symbols (such as machined or unmachined) in a drawing file.

Symbol

This command is used to create and manipulate symbols within a drawing file. Symbols are collections of many draft elements and/or text. You may create and use *simple* symbols (exact duplicates of the original) or *generic* symbols (variations on the original), which are similar to a family of parts.

X-section

This command is used to apply a cross-hatch or filled pattern to a cross-section view, using the various options available.

DRAWING ➥ RETRIEVE ➥ DETAIL ➥ CREATE ➥ NOTE

DETAIL
Show
Erase
Create
Delete
Move
Move Text
Move Attach
Break
Clip
Switch View
Flip Arrows
Make Job
Sketch
Tools
Modify
Done / Return

DETAIL ITEM
Dimension
Ref Dim
Note
Geom Tol
Surf Finish
Datum
X-section
Balloon
Axis
Symbol

NOTE TYPES
No Leader
Leader
On Item
Enter
File
Horizontal
Vertical
Angular
Standard
Normal Ldr
Tangent Ldr
Left
Center
Right
Default
Make Note
Done / Return

Angular

This command is used to place a note that is neither horizontal nor vertical in its orientation. You will be prompted to enter a positive angle value between 0 — 359 degrees.

Center

This command is used to place a note using justification, so that the center of the note will be located at the specified point.

Default

This command is used to place a note using justification, so that the left side of the note will be located at the specified point (if the note is free). If the note has a leader line attached to it, this command instructs Pro/ENGINEER to justify the text so that the side with the leader is the active justification.

Enter

This command is used to place a note, by entering the content via the keyboard.

File

This command is used to place a note, by specifying a text file containing the note's contents.

Horizontal

This command is used to place a note that is horizontal in its orientation.

Leader

This command is used to place a note that is attached to a model edge or draft entity. Notes with leader lines may also be placed in an unattached manner by using the **Free Point** option.

Left

This command is used to place a note using justification so that the left side of the note will be located at the specified point.

Make Note

This command is used after all the note parameters have been established (type, justification, leader type, etc.) and initiates the creation of the drawing note.

No Leader

This command is used to place a *free standing* note that is not attached to a model edge via a leader line.

Normal Ldr

This command is used to place a note with a leader line that is normal to a selected entity.

On Item

This command is used to place a *free standing* note that ***is*** attached to a model edge or datum point.

Right

This command is used to place text using justification so that the right side of the text will be located at the specified point.

Standard

This command is used to place a note with a leader line that is located by the points you specify.

Tangent Ldr

This command is used to place a note with a leader line that is tangent to a selected entity.

Vertical

This command is used to place a note that is vertical in its orientation.

DRAWING ➥ RETRIEVE ➥ DETAIL ➥ CREATE ➥ NOTE ➥ LEADER

DETAIL ITEM
Dimension
Ref Dim
Note
Geom Tol
Surf Finish
Datum
X-section
Balloon
Axis
Symbol

NOTE TYPES
No Leader
Leader
On Item
Enter
File
Horizontal
Vertical
Angular
Standard
Normal Ldr
Tangent Ldr
Left
Center
Right
Default
Make Note
Done / Return

ATTACH TYPE
On Entity
Free Point
Midpoint
Intersect
Arrow Head
Dot
Filled Dot
No Arrow
Done
Quit

Arrow Head

This command is used to terminate the note's leader line with a filled arrow head.

Dot

This command is used to terminate the note's leader line with a hollow circle.

Filled Dot

This command is used to terminate the note's leader line with a filled circle.

Free Point

This command is used to place a note with a leader line (anywhere in a drawing file) that is not attached to a model edge or draft entity.

Intersect

This command is used to place a note with a leader line that is attached to the intersection of two model edges or draft entities.

Midpoint

This command is used to place a note with a leader line that is attached to the midpoint of a model edge or draft entity.

No Arrow

This command is used to place a note with a non-terminated leader line.

On Entity

This command is used to place a note with a leader line that is attached to the vertex of a model edge or draft entity.

DRAWING ➡ RETRIEVE ➡ DETAIL ➡ CREATE ➡ SURF FINISH

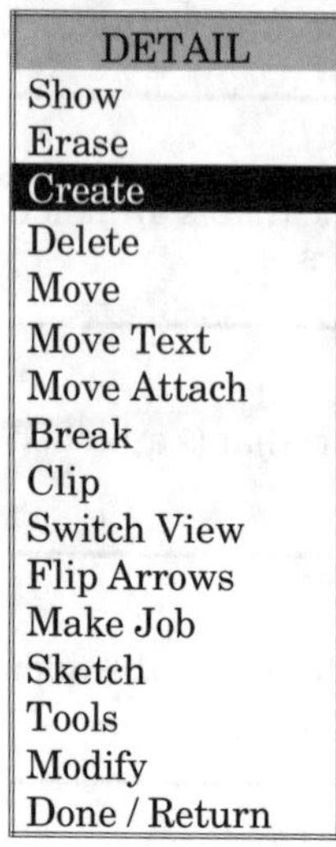

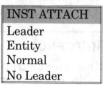

DETAIL ITEM
Dimension
Ref Dim
Note
Geom Tol
Surf Finish
Datum
X-section
Balloon
Axis
Symbol

INST ATTACH
Leader
Entity
Normal
No Leader

Entity

This command is used to attach a surface finish symbol to a model edge or draft element.

Leader

This command is used to place a surface finish symbol with a leader line.

Name

This command is used to recall and place a surface finish symbol by selecting its name from a displayed list. The names of all the symbols currently in the drawing file will be displayed in the menu.

No Leader

This command is used to place a surface finish symbol with no leader line.

Normal

This command is used to attach a surface finish symbol to a model edge or draft element with the vertical reference pointing up. You will be required to define "UP" when attaching to an angled item.

Pick Inst

This command is used to place a surface finish symbol. The type of finish symbol is determined by selecting any instance of the desired symbol in the drawing file.

Detail - Erase

DRAWING ➡ RETRIEVE ➡ DETAIL ➡ ERASE

DRAWING
Views
Sheets
Modify
Regenerate
Switch Dim
Relations
Detail
Interface
Dwg Format
Table
Layer
Set Up
User Attrbt
Symbol
Set Model
Info
Represent
Clean Dims

DETAIL
Show
Erase
Create
Delete
Move
Move Text
Move Attach
Break
Clip
Switch View
Flip Arrows
Make Job
Sketch
Tools
Modify
Done / Return

ERASE ITEM
Erase All
Feat & View
By Feat
By View
One Item
Dimension
Ref Dim
Note
Geom Tol
Surf Finish
Datum
X-section
Balloon
Axis
Symbol

By Feature

This command is used to erase the specified information about a selected feature.

By View

This command is used to erase all the detail items associated with a specified view.

Erase All

This command is used to erase all drawing items. You will be prompted to confirm this command.

Feat & View

This command is used to erase the specified information about a selected feature, but only in a specified view.

One Item

This command is used to erase a single drawing item, regardless of the view it is associated with.

Detail - Modify

DRAWING ➡ RETRIEVE ➡ DETAIL ➡ MODIFY ➡ DIM PARAMS

DETAIL
Show
Erase
Create
Delete
Move
Move Text
Move Attach
Break
Clip
Switch View
Flip Arrows
Make Job
Sketch
Tools
Modify
Done / Return

MODIFY DRAW
Value
Num Digits
Text
Xhatching
Grid
Symbol
Line Style
Geom Tol
Dim Params
Diameter
Done / Return

DIM PARAMS
Skew
Format
Dim Symbol
Dim Type
Scheme
Return

Dim Symbol

This command is used to replace the default dimension tag created by Pro/ENGINEER (d0,d1) with a user specified tag (SIDE_A).

Dim Type

This command is used to choose a dimensioning type other than **standard**.

Skew

This command is used to modify the witness lines of linear dimensions. It is similar in theory to the **Slant** option which may be applied to text elements.

DRAWING ➥ RETRIEVE ➥ DETAIL ➥ MODIFY ➥ DIM PARAMS ➥ DIM TYPE

DETAIL
Show
Erase
Create
Delete
Move
Move Text
Move Attach
Break
Clip
Switch View
Flip Arrows
Make Job
Sketch
Tools
Modify
Done / Return

MODIFY DRAW
Value
Num Digits
Text
Xhatching
Grid
Symbol
Line Style
Geom Tol
Dim Params
Diameter
Done / Return

DIM PARAMS
Skew
Format
Dim Symbol
Dim Type
Scheme
Return

DIM TYPE
Coord Dim
Ordinate Dim
Quit

Coordinate Dim

This command instructs Pro/ENGINEER that subsequent dimensions placed in the drawing file will be displayed using the coordinate dimension type (X=, Y=).

Standard Dim

This command instructs Pro/ENGINEER that subsequent dimensions placed in the drawing file will be displayed using the **standard** dimension type.

Ordinate Dim

This command instructs Pro/ENGINEER that subsequent dimensions placed in the drawing file will be displayed using the ordinate dimension type (relative distance from a known **base**).

DRAWING ➥ RETRIEVE ➥ DETAIL ➥ MODIFY ➥ DIM PARAMS ➥ ORDINATE DIM

DETAIL
Show
Erase
Create
Delete
Move
Move Text
Move Attach
Break
Clip
Switch View
Flip Arrows
Make Job
Sketch
Tools
Modify
Done / Return

MODIFY DRAW
Value
Num Digits
Text
Xhatching
Grid
Symbol
Line Style
Geom Tol
Dim Params
Diameter
Done / Return

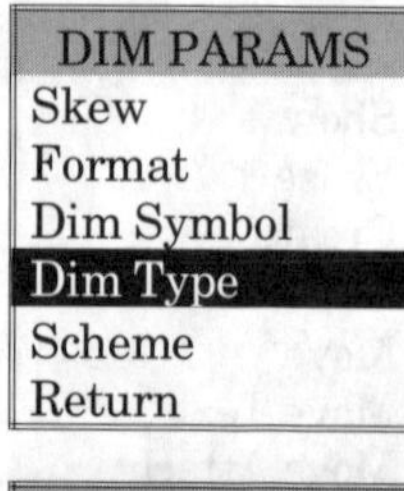

DIM PARAMS
Skew
Format
Dim Symbol
Dim Type
Scheme
Return

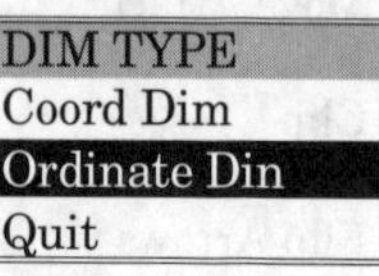

DIM TYPE
Coord Dim
Ordinate Din
Quit

MOD DIM TYPE
Create Base
Set Base
Lin to Ord
Ord to Lin
Done / Quit

Create Base

This command is used to convert a linear dimension into an ordinate dimension. The dimension selected will become the **base** for all other ordinate dimensions placed in the drawing file.

Lin to Ord

This command is used to convert a linear dimension into an ordinate dimension (after a **base** has been established).

Ord to Lin

This command is used to convert an ordinate dimension into a linear dimension.

Set Base

This command is used to set a baseline reference when converting a linear dimension to an ordinate dimension.

DRAWING ➥ RETRIEVE ➥ DETAIL ➥ MODIFY ➥ GRID

DETAIL
Show
Erase
Create
Delete
Move
Move Text
Move Attach
Break
Clip
Switch View
Flip Arrows
Make Job
Sketch
Tools
Modify
Done / Return

MODIFY DRAW
Value
Num Digits
Text
Xhatching
Grid
Symbol
Line Style
Geom Tol
Dim Params
Diameter
Done / Return

GRID MODIFY
Grid On
Grid Off
Type
Origin
Grid Params

Grid Off

Turns **off** the grid display. As with **Sketcher**, this command has no effect on the *snapping* capability of the grid. This is determined by the value set for **Grd Snp** under **Environment**.

Grid On

Turns **on** the grid display. As with **Sketcher**, this command has no effect on the *snapping* capability of the grid. This is determined by the value set for **Grd Snp** under **Environment**.

Grid Params

Allows you to specify the spacing and angles of various grid variables for both Cartesian and Polar grids.

Origin

Allows you to relocate the grid's intersection point. The acceptable locations are sketched points, datum points, vertices of curves or edges, and the endpoint or midpoint of a draft element.

Type

Allows you to specify a grid type of **Cartesian** or **Polar**.

DRAWING ➥ RETRIEVE ➥ DETAIL ➥ MODIFY ➥ GRID ➥ TYPE

DETAIL
Show
Erase
Create
Delete
Move
Move Text
Move Attach
Break
Clip
Switch View
Flip Arrows
Make Job
Sketch
Tools
Modify
Done / Return

MODIFY DRAW
Value
Num Digits
Text
Xhatching
Grid
Symbol
Line Style
Geom Tol
Dim Params
Diameter
Done / Return

GRID MODIFY
Grid On
Grid Off
Type
Origin
Grid Params

GRID TYPE
Cartesian
Polar
Done / Return

Cartesian

Instructs Pro/ENGINEER to display a **Cartesian** type of rectangular grid.

Polar

Instructs Pro/ENGINEER to display a **Polar** type of radial grid.

DRAWING ➥ RETRIEVE ➥ DETAIL ➥ MODIFY ➥ GRID ➥ TYPE ➥ CARTESIAN

DETAIL
Show
Erase
Create
Delete
Move
Move Text
Move Attach
Break
Clip
Switch View
Flip Arrows
Make Job
Sketch
Tools
Modify
Done / Return

MODIFY DRAW
Value
Num Digits
Text
Xhatching
Grid
Symbol
Line Style
Geom Tol
Dim Params
Diameter
Done / Return

GRID MODIFY
Grid On
Grid Off
Type
Origin
Grid Params

CART PARAMS
X&Y Spacing
X Spacing
Y Spacing
Angle
Done / Return

Angle

Specifies the angle between horizontal and the X grid axis (a rotated grid pattern).

X Spacing

Sets the spacing for the X grid value only.

X&Y Spacing

Simultaneously sets the X and Y grid spacing to a specified value.

Y Spacing

Sets the spacing for the Y grid value only.

DRAWING ➡ RETRIEVE ➡ DETAIL ➡ MODIFY ➡ GRID ➡ TYPE ➡ POLAR

DETAIL
Show
Erase
Create
Delete
Move
Move Text
Move Attach
Break
Clip
Switch View
Flip Arrows
Make Job
Sketch
Tools
Modify
Done / Return

MODIFY DRAW
Value
Num Digits
Text
Xhatching
Grid
Symbol
Line Style
Geom Tol
Dim Params
Diameter
Done / Return

GRID MODIFY
Grid On
Grid Off
Type
Origin
Grid Params

POLAR PARAMS
Ang Spacing
Num Lines
Rad Spacing
Angle
Done / Return

Angle

Specifies the angle between horizontal and the 0-degree grid axis (a rotated grid pattern).

Ang Spacing

Specifies the angular spacing between radial lines. The value must divide evenly into 360.

Num Lines

Specifies the quantity of radial lines in the grid pattern based upon the formula (360/number of lines).

Rad Spacing

Specifies the spacing of the circular grid indicators.

DRAWING ➥ RETRIEVE ➥ DETAIL ➥ MODIFY ➥ LINE STYLE

MODIFY DRAW
Value
Num Digits
Text
Xhatching
Grid
Symbol
Line Style
Geom Tol
Dim Params
Diameter
Done / Return

CROSS XHATCH
Spacing
Angle
Offset
Line Style
Add Line
Delete Line
Next Line
Prev Line
Retrieve
Save
Hatch
Fill
Excl Comp
Restore Comp
Next Xsec
Prev Xsec
Done
Quit

LINE STYLE
Geometry
Hidden
Leader
Cut Plane
Phantom
Centerline
Other
Done / Quit

Centerline

This command is used to modify the display of cosmetic/draft geometry so that it appears as **yellow-centerline** on the screen and plots as centerlines.

Cut Plane

This command is used to modify the display of cosmetic/draft geometry so that it appears as **white-phantom** on the screen.

Geometry

This command is used to modify the display of cosmetic/draft geometry so that it appears as **white** on the screen (default) and plots as a solid lines.

Hidden

This command is used to modify the display of cosmetic/draft geometry so that it appears as **gray** on the screen and plots as a dashed lines.

Leader

This command is used to modify the display of cosmetic/draft geometry so that it appears as **yellow** on the screen.

Other

This command displays a sub-menu allowing you to specify the color, style, and thickness that you wish the cosmetic/draft geometry to appear.

Phantom

This command is used to modify the display of cosmetic/draft geometry so that it appears as **gray-phantom** on the screen and plots as phantom lines.

DRAWING ➡ RETRIEVE ➡ DETAIL ➡ MODIFY ➡ TEXT

DETAIL
Show
Erase
Create
Delete
Move
Move Text
Move Attach
Break
Clip
Switch View
Flip Arrows
Make Job
Sketch
Tools
Modify
Done / Return

MODIFY DRAW
Value
Num Digits
Text
Xhatching
Grid
Symbol
Line Style
Geom Tol
Dim Params
Diameter
Done / Return

MODIFY TEXT
Text Line
Full Note
Justify
Angle
Text Height
Text Width
Slant Angle
Thickness
Font
Underlining
Mirror
Done / Return

Angle

This command is used to rotate a note by a specified angle.

Font

This command is used to modify the text character font.

Full Note

This command is used to modify many lines of text at one time. The note will be loaded into a editor window. Using this command, you may modify any part of the note.

Justify

This command is used to modify the justification of a note (center, left, right, etc.).

Mirror

This command is used to mirror a selected note about a draft line.

Slant Angle

This command is used to apply a slant angle (italicize) to text characters.

Text Height

This command is used to modify the height of text characters.

Text Line

This command is used to modify a *single* text string, or to add a text string to a dimension.

Text Width

This command is used to modify the width of text characters by specifying a ratio that is applied to the character's height. Acceptable values range from 0.25 to 8 (example: .5 is equal to 50 percent of the text height).

Thickness

This command is used to apply a thickness to text characters, making them appear heavier. Acceptable values range from zero - five (zero is the default).

Underlining

This command is used to apply or remove an underlined characteristic from a selected note.

DRAWING ➡ RETRIEVE ➡ DETAIL ➡ MODIFY ➡ XHATCHING

DETAIL
Show
Erase
Create
Delete
Move
Move Text
Move Attach
Break
Clip
Switch View
Flip Arrows
Make Job
Sketch
Tools
Modify
Done / Return

MODIFY DRAW
Value
Num Digits
Text
Xhatching
Grid
Symbol
Line Style
Geom Tol
Dim Params
Diameter
Done / Return

CROSS XHATCH
Spacing
Angle
Offset
Line Style
Add Line
Delete Line
Next Line
Prev Line
Retrieve
Save
Hatch
Fill
Excl Comp
Restore Comp
Next Xsec
Prev Xsec
Done
Quit

Angle

This command is used to modify the angle of one or many sets of cross-hatch pattern lines.

Excl Comp

This command is used to exclude an assembly component from a cross-section display.

Fill

This command is used to apply a *filled pattern* to a cross-section display, rather than a filled pattern.

Hatch

This command is used to apply a *cross-hatch* pattern to a cross-section display, rather than a cross-hatch pattern.

Line Style

This command is used to modify the appearance of cross-hatch pattern lines.

Next Xsec

This command is used when modifying a cross-hatch pattern that is part of an assembly. This command changes the currently highlighted cross-hatch pattern.

Prev Xsec

This command is used when modifying a cross-hatch pattern that is part of an assembly. This command changes the currently highlighted cross-hatch pattern.

Restore Comp

This command is used to restore an assembly component that has been previously excluded using the **Excl Comp** command.

Spacing

This command is used to modify the spacing between cross-hatch pattern lines.

DRAWING ➥ RETRIEVE ➥ DETAIL ➥ MODIFY ➥ XHATCHING ➥ ANGLE

MODIFY DRAW
Value
Num Digits
Text
Xhatching
Grid
Symbol
Line Style
Geom Tol
Dim Params
Diameter
Done / Return

CROSS XHATCH
Spacing
Angle
Offset
Line Style
Add Line
Delete Line
Next Line
Prev Line
Retrieve
Save
Hatch
Fill
Excl Comp
Restore Comp
Next Xsec
Prev Xsec
Done
Quit

MODIFY MODE
Individual
Overall
0
30
45
60
90
120
135
150
Value

Individual

This command instructs Pro/ENGINEER to apply the modifications to only one set of cross-hatch pattern lines.

Overall

This command instructs Pro/ENGINEER to apply the modifications to both sets of cross-hatch pattern lines.

Value

This command is used to specify an angle other than those supplied on the menu.

DRAWING ➡ RETRIEVE ➡ DETAIL ➡ MODIFY ➡ XHATCHING ➡ LINE STYLE

MODIFY DRAW	CROSS XHATCH	LINE STYLE
Value	Spacing	Geometry
Num Digits	Angle	Hidden
Text	Offset	Leader
Xhatching	Line Style	Cut Plane
Grid	Add Line	Phantom
Symbol	Delete Line	Centerline
Line Style	Next Line	Other
Geom Tol	Prev Line	Done / Quit
Dim Params	Retrieve	
Diameter	Save	
Done / Return	Hatch	
	Fill	
	Excl Comp	
	Restore Comp	
	Next Xsec	
	Prev Xsec	
	Done	
	Quit	

Centerline

This command is used to modify the display of cross-hatch pattern lines so they appear as **yellow-centerline** on the screen and plot as centerlines.

Cut Plane

This command is used to modify the display of cross-hatch pattern lines so they appear as **white-phantom** on the screen.

Geometry

This command is used to modify the display of cross-hatch pattern lines so they appear as **white** on the screen (default) and plot as solid lines.

Hidden

This command is used to modify the display of cross-hatch pattern lines so they appear as **gray** on the screen and plot as dashed lines.

Leader

This command is used to modify the display of cross-hatch pattern lines so they appear as **yellow** on the screen.

Other

This command displays a sub-menu allowing you to specify the color, style, and thickness that you wish the cross-hatch pattern lines to appear.

Phantom

This command is used to modify the display of cross-hatch pattern lines so they appear as **gray-phantom** on the screen and plot as phantom lines.

DRAWING ➥ RETRIEVE ➥ DETAIL ➥ MODIFY ➥ XHATCHING ➥ SPACING

MODIFY DRAW
Value
Num Digits
Text
Xhatching
Grid
Symbol
Line Style
Geom Tol
Dim Params
Diameter
Done / Return

CROSS XHATCH
Spacing
Angle
Offset
Line Style
Add Line
Delete Line
Next Line
Prev Line
Retrieve
Save
Hatch
Fill
Excl Comp
Restore Comp
Next Xsec
Prev Xsec
Done
Quit

MODIFY MODE
Individual
Overall
Half
Double
Value

Double

This command is used to increase the current spacing between cross-hatch pattern lines by 200 percent.

Half

This command is used to decrease the current spacing between cross-hatch pattern lines by 50 percent.

Detail - Move Attach

DRAWING ➥ RETRIEVE ➥ DETAIL ➥ MOVE ATTACH

DRAWING
Views
Sheets
Modify
Regenerate
Switch Dim
Relations
Detail
Interface
Dwg Format
Table
Layer
Set Up
User Attrbt
Symbol
Set Model
Info
Represent
Clean Dims

DETAIL
Show
Erase
Create
Delete
Move
Move Text
Move Attach
Break
Clip
Switch View
Flip Arrows
Make Job
Sketch
Tools
Modify
Done / Return

Same Ref
Change Ref

Change Ref

This command is used to relocate the leader line's attachment point to another element.

Same Ref

This command is used to relocate the leader line's attachment point to another location on the same element.

Detail - Show

DRAWING ➥ RETRIEVE ➥ DETAIL ➥ SHOW

DETAIL
Show
Erase
Create
Delete
Move
Move Text
Move Attach
Break
Clip
Switch View
Flip Arrows
Make Job
Sketch
Tools
Modify
Done / Return

DETAIL ITEM
Dimension
Ref Dim
Note
Geom Tol
Surf Finish
Dautm
X-section
Balloon
Axis
Symbol

SHOW ITEM
Show All
By View
By Feature
Feature & View

NOTE: Selecting any item, except Dimension, from the Detail Item menu brings up the Show Item menu.

By Feature

This command is used to display the specified information about a selected feature, in all views.

By View

This command is used to display all the specified information about a part, but only in a selected view.

Feat & View

This command is used to display the specified information about a feature, but only in a selected view.

Show All

This command is used to display all the specified information about a part, in all the views.

Detail - Sketch

DRAWING ➥ RETRIEVE ➥ DETAIL ➥ SKETCH

DRAWING	DETAIL	DRAFT GEOM
Views	Show	Line
Sheets	Erase	Circle
Modify	Create	Arc
Regenerate	Delete	Construction
Switch Dim	Move	Other
Relations	Move Text	Return
Detail	Move Attach	Start Chain
Interface	Break	End Chain
Dwg Format	Clip	
Table	Switch View	
Layer	Flip Arrows	
Set Up	Make Job	
User Attrbt	**Sketch**	
Symbol	Tools	
Set Model	Modify	
Info	Done / Return	
Represent		

Arc

This command is used to create an arc using one of the various placement options.

Circle

This command is used to create a circle using one of the various placement options.

Construction

This command is used to create a *construction* class, line, or circle using one of the various placement options. Construction class elements are typically used as temporary geometry to assist in the creation of draft geometry.

End Chain

This command is used to terminate the *chaining* process started by the **Start Chain** command.

Line

This command is used to create a line using one of the various placement options.

Other

This command is used to create a spline, ellipse, point, or chamfer, using the various placement options.

Start Chain

This command is used to create a series of new elements by picking points upon existing geometry.

DRAWING ➥ RETRIEVE ➥ DETAIL ➥ SKETCH ➥ ARC

DETAIL
Show
Erase
Create
Delete
Move
Move Text
Move Attach
Break
Clip
Switch View
Flip Arrows
Make Job
Sketch
Tools
Modify
Done / Return

DRAFT GEOM
Line
Circle
Arc
Construction
Other
Return
Start Chain
End Chain

ARC
Tang End
3 Points
Ctr / Ends
Pnt / Ctr / Ang
3 Tangent
Fillet

3 Points

This command is used to create an arc by first locating it's beginning and ending points, then specifying a third point that controls the radius of the arc.

3 Tangent

This command is used to create an arc by first selecting three existing elements that the arc will be tangent to. The order in which you select the elements will control the placement of the new arc.

Ctr / Ends

This command is used to create an arc by first locating a point on the arc (which becomes the arc's first endpoint), then locating the arc's center, finally specifying a third point which controls the arc's other endpoint. You will see a two color circle and you must specify which colored arc you wish to keep.

Fillet

This command is used to create an arc tangent to two existing elements, by selecting the existing elements and then specifying the arc's radius.

Pnt / Ctr / Ang

This command is used to create an arc by selecting a starting point and a center point (which define the first axis of the arc element). The other axis (and endpoint) will be determined by a specified sweep angle and a counter-clockwise orientation.

Tang End

This command is used to create an arc in which the first endpoint is *tangent* to a selected element.

DRAWING ➥ RETRIEVE ➥ DETAIL ➥ SKETCH ➥ CIRCLE

DETAIL
Show
Erase
Create
Delete
Move
Move Text
Move Attach
Break
Clip
Switch View
Flip Arrows
Make Job
Sketch
Tools
Modify
Done / Return

DRAFT GEOM
Line
Circle
Arc
Construction
Other
Return
Start Chain
End Chain

CIRCLE
Center / Pnt
Center / Tang
Center / Rad
Center / Dia
3 Tangent
Fillet
3 Points

3 Points

This command is used to create a circle in which the edge will be located on any three specified points.

3 Tangent

This command is used to create a circle by selecting three elements to which the circle's edge will be tangent.

Center / Dia

This command is used to create a circle by selecting its center point and entering its diameter via the keyboard.

Center / Pnt

This command is used to create a circle by selecting its center point and an arbitrary radius point.

Center / Rad

This command is used to create a circle by selecting its center point and entering its radius via the keyboard.

Center / Tang

This command is used to create a circle by selecting its center point and an existing element to which the circle's edge will be tangent.

Fillet

This command is used to create a circle that is tangent to two existing elements, and has a specified radius.

DRAWING ➥ RETRIEVE ➥ DETAIL ➥ SKETCH ➥ CONSTRUCTION

DETAIL
Show
Erase
Create
Delete
Move
Move Text
Move Attach
Break
Clip
Switch View
Flip Arrows
Make Job
Sketch
Tools
Modify
Done / Return

DRAFT GEOM
Line
Circle
Arc
Construction
Other
Return
Start Chain
End Chain

CONSTRUCT
Constr Line
Constr Circ

Constr Circ

This command is used to create a construction class circle.

Constr Line

This command is used to create a construction class line.

DRAWING ➥ RETRIEVE ➥ DETAIL ➥ SKETCH ➥ LINE

DETAIL
Show
Erase
Create
Delete
Move
Move Text
Move Attach
Break
Clip
Switch View
Flip Arrows
Make Job
Sketch
Tools
Modify
Done / Return

DRAFT GEOM
Line
Circle
Arc
Construction
Other
Return
Start Chain
End Chain

LINE
2 Points
Horiz Line
Vert Line
Angle
Pnt / Tang
Parl Line
Perp Line
Norm At Pnt
Tang Line
2 Tang Line

2 Points

This command is used to create a line between any two selected points.

2 Tang Line

This command is used to create a line that is tangent to two selected arcs, circles, or splines.

Angle

This command is used to create a line at a specified angle from the X axis.

Horiz Line

This command is used to create a horizontal line (along the X axis) between any two selected points.

Norm At Pnt

This command is used to create a line that is normal to a selected curve, and is terminated by a selected element. Select the object to be normal to (the selection point is the line's first endpoint), then select a point on another element whose intersection with the new line will determine the new line's length.

Parl Line

This command is used to create a line that is parallel to another, by selecting the line's starting point, selecting an existing line to be parallel to, and then picking the new line's ending point.

Perp Line

This command is used to create a line that is perpendicular to another by selecting the new line's starting point, selecting an existing line to be perpendicular to, then picking the new line's ending point.

Pnt / Tang

This command is used to create a line that is tangent to an existing curve at its second endpoint.

Tang Line

This command is used to create a line that is tangent to an existing curve at its first endpoint.

Vert Line

This command is used to create a vertical line (along the Y axis) between any two selected points.

DRAWING ➥ RETRIEVE ➥ DETAIL ➥ SKETCH ➥ OTHER

DETAIL
Show
Erase
Create
Delete
Move
Move Text
Move Attach
Break
Clip
Switch View
Flip Arrows
Make Job
Sketch
Tools
Modify
Done / Return

DRAFT GEOM
Line
Circle
Arc
Construction
Other
Return
Start Chain
End Chain

OTHER GEOM
Spline
Ellipse
Point
Chamfer
Return
Start Chain
End Chain

Chamfer

This command is used to create a line which intersects and trims two existing lines, at specified distances from the intersection of the two original lines. The original lines do not have to physically intersect on the screen, but their trajectory must eventually come to an intersection.

Ellipse

This command is used to create an ellipse using the various options available.

Point

This command is used to create a point to be used as a reference for locating other geometry.

Spline

This command is used to create a B-spline element that will traverse through many selected points. To end the placement of a spline element, press the middle mouse button.

DRAWING ➥ RETRIEVE ➥ DETAIL ➥ SKETCH ➥ OTHER ➥ CHAMFER

DETAIL
- Show
- Erase
- Create
- Delete
- Move
- Move Text
- Move Attach
- Break
- Clip
- Switch View
- Flip Arrows
- Make Job
- Sketch
- Tools
- Modify
- Done / Return

DRAFT GEOM
- Line
- Circle
- Arc
- Construction
- Other
- Return
- Start Chain
- End Chain

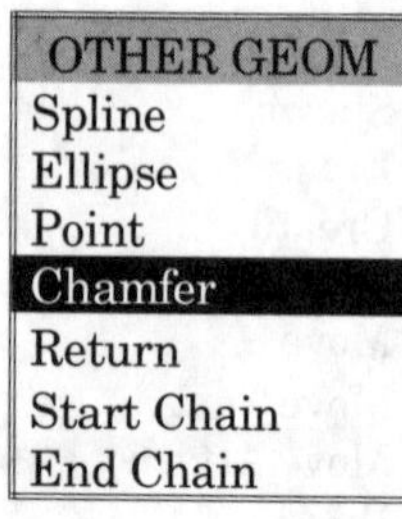

CHAMFER
- 45 x d
- d x d
- d1 x d2
- Ang x d
- Quit

45 x d

This command is used to create a chamfer that is located at a 45 degree angle from the first line selected, and at a specified distance (d) from the intersection along both lines.

Ang x d

This command is used to create a chamfer that is located at an angle, and at a specified distance (d) from the intersection of the two existing lines.

d x d

This command is used to locate a chamfer that is located at an equal distance (d) from the intersection of the two existing lines.

d1 x d2

This command is used to create a chamfer that is located at different distances from the intersection of the two existing lines.

DRAWING ➡ RETRIEVE ➡ DETAIL ➡ SKETCH ➡ OTHER ➡ ELLIPSE

DETAIL
Show
Erase
Create
Delete
Move
Move Text
Move Attach
Break
Clip
Switch View
Flip Arrows
Make Job
Sketch
Tools
Modify
Done / Return

DRAFT GEOM
Line
Circle
Arc
Construction
Other
Return
Start Chain
End Chain

OTHER GEOM
Spline
Ellipse
Point
Chamfer
Return
Start Chain
End Chain

ELLIPSE
Ctr / Pnt / Ang
Pnt / Pnt / Ang
Ctr / Pnt / Pnt
3 Points

3 Points

This command is used to create an ellipse by selecting one of the ellipse's edge points, selecting another point that will establish the major axis of the ellipse and its length, then selecting a third point that will define the minor axis and it's length.

Ctr / Pnt / Ang

This command is used to create an ellipse by selecting the ellipse's center point, selecting another point that will establish the major axis of the ellipse and half its length, then entering an angle, via the keyboard, at which the ellipse will rotate about the major axis.

Ctr / Pnt / Pnt

This command is used to create an ellipse by selecting the ellipse's center point, electing another point that will establish the major axis of the ellipse and half its length, then selecting a third point that will define the minor axis and it's length.

Pnt / Pnt / Ang

This command is used to create an ellipse by selecting one of the ellipse's edge points, electing another point that will establish the major axis of the ellipse and its length, then entering an angle, via the keyboard, at which the ellipse will rotate about the major axis.

Detail - Tools

DRAWING ➥ RETRIEVE ➥ DETAIL ➥ TOOLS

DRAWING
Views
Sheets
Modify
Regenerate
Switch Dim
Relations
Detail
Interface
Dwg Format
Table
Layer
Set Up
User Attrbt
Symbol
Set Model
Info
Represent

DETAIL
Show
Erase
Create
Delete
Move
Move Text
Move Attach
Break
Clip
Switch View
Flip Arrows
Make Job
Sketch
Tools
Modify
Done / Return

TOOLS
Translate
Rotate
Rescale
Copy
Mirror
Trim
Intersect
Group
Offset
Use Edge
Define Char
Draft View
Done / Return

Copy

This command is used to duplicate existing draft elements by translating them along an X and/or Y axis, or by rotating them about a specified point.

Group

This command is used to group many draft elements together so they may be manipulated as one.

Intersect

This command is used to find the intersection of two existing elements and break both elements at that point.

Mirror

This command is used to duplicate existing draft elements about a draft line.

Offset

This command is used to create draft elements using the various options to offset them from existing geometry.

Rescale

This command is used to modify the overall size of a draft element.

Trim

This command is used to modify the length of an existing element by using the available options. Unlike with the **Intertsect** command, no new elements are created.

Use Edge

This command is used to create draft elements by extracting their endpoints and orientation from a model edge.

DRAWING ➡ RETRIEVE ➡ DETAIL ➡ TOOLS ➡ COPY ➡ TRANSLATE

DETAIL
Show
Erase
Create
Delete
Move
Move Text
Move Attach
Break
Clip
Switch View
Flip Arrows
Make Job
Sketch
Tools
Modify
Done / Return

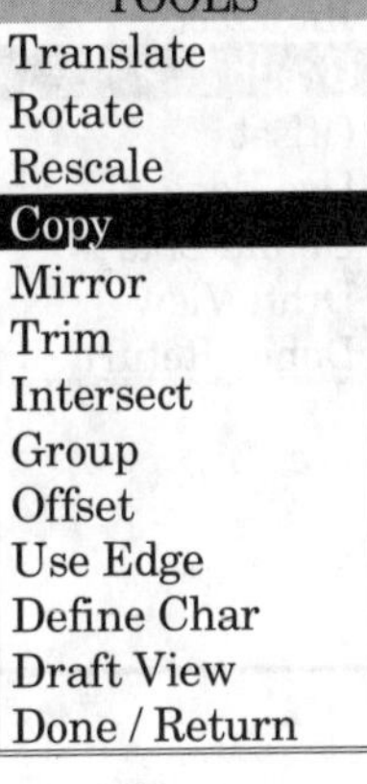

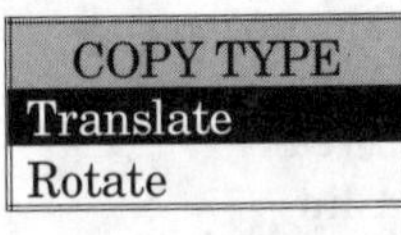

GET VECTOR
Horiz
Vert
Ang / Length
From - To

Ang/Length

Translates elements by a specified distance along a vector. The vector is defined by an angle measured from the X axis in a counter-clockwise rotation.

From — To

Translates elements along a vector defined by two arbitrarily selected points.

Horiz

Translates elements along the X axis using specified units.

Vert

Translates elements along the Y axis using specified units.

DRAWING ➡ RETRIEVE ➡ DETAIL ➡ TOOLS ➡ GROUP

DETAIL
Show
Erase
Create
Delete
Move
Move Text
Move Attach
Break
Clip
Switch View
Flip Arrows
Make Job
Sketch
Tools
Modify
Done / Return

TOOLS
Translate
Rotate
Rescale
Copy
Mirror
Trim
Intersect
Group
Offset
Use Edge
Define Char
Draft View
Done / Return

DRAFT GROUP
Create
Suppress
Resume
Explode
Edit

GROUP ACCESS
Select
By Name

By Name

This command is used to specify which groups will be suppressed or exploded by entering the group's name.

Create

This command is used to create a new draft group, by selecting the members and assigning a name to the group.

Edit

This command is used to **add** or **remove** a draft entity from a group.

Explode

This command is used to delete the grouping that associates many items, without removing the items from the drawing file.

Resume

This command is used to reverse the effects of the **Suppress** command.

Select

This command is used to specify which groups will be suppressed or exploded by selecting the group visually.

Suppress

The command is used to remove a group's members from the current display.

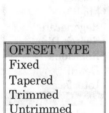

DRAWING ➡ RETRIEVE ➡ DETAIL ➡ TOOLS ➡ OFFSET

DETAIL
Show
Erase
Create
Delete
Move
Move Text
Move Attach
Break
Clip
Switch View
Flip Arrows
Make Job
Sketch
Tools
Modify
Done / Return

TOOLS
Translate
Rotate
Rescale
Copy
Mirror
Trim
Intersect
Group
Offset
Use Edge
Define Char
Draft View
Done / Return

OFFSET OPER
Single Ent
Ent Chain

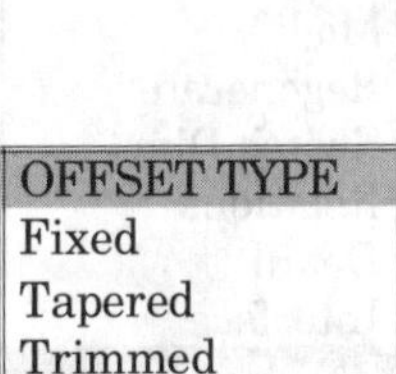

OFFSET TYPE
Fixed
Tapered
Trimmed
Untrimmed

Ent Chain

This command is used to offset many elements at a time. The selected elements must be connected.

Fixed

This command is used to create a draft element that is offset by a single value (parallel) to an existing element.

Single Ent

This command is used to offset one element at a time.

Tapered

This command is used to create a draft element that is offset by a unique value at each endpoint.

Trimmed

This command instructs Pro/ENGINEER to offset only a portion of the selected draft element using the specified options.

Untrimmed

This command instructs Pro/ENGINEER to offset the entire draft element using the specified options.

Drawing Format

DRAWING ➡ RETRIEVE ➡ DWG FORMAT

DRAWING
Views
Sheets
Modify
Regenerate
Switch Dim
Relations
Detail
Interface
Dwg Format
Table
Layer
Set Up
User Attrbt
Symbol
Set Model
Info
Represent

DRAWFORMAT
List
Add / Replace
Remove

Add/Replace

This command is used to select an existing drawing format to be associated to the current drawing file. If the drawing format is already associated with the drawing file, it will be replaced by the new selection.

Remove

This command is used to remove any association between a drawing format file and a drawing file which may already exist.

List

This command is used to display a list of drawing format files you may attach to your drawing file.

Info

DRAWING ➥ RETRIEVE ➥ INFO

DRAWING
Views
Sheets
Modify
Regenerate
Switch Dim
Relations
Detail
Interface
Dwg Format
Table
Layer
Set Up
User Attrbt
Symbol
Set Model
Info
Represent

INFO
Mass Props
Names
Measure
Feat Info
Model Info
Comp Info
ParentChild
Layer Info
Drawing

Comp Info

This command is used to display information regarding the process used to assemble a selected component.

Drawing

This command displays a sub-menu that allows you to access information about specific drawing entities.

Feat Info

This command displays various types of information (name, units, dimensions, feature ID, etc.) for a selected feature.

Layer Info

This command is used to display information about the layers defined in a specified drawing or part.

Mass Prop

This command displays a sub-menu that allows you to calculate the mass properties of a part or cross-section associated to your drawing.

Model Info

This command displays various types of information (name, units, dimensions, feature ID, etc.) for a selected part.

Names

This command is used to display a list of all the files in the current directory, regardless of type (part, assembly, drawing, layout, sections, etc.).

Parent / Child

This command is used to display information about the parent / child relationships between features.

DRAWING ➥ RETRIEVE ➥ INFO ➥ DRAWING

DRAWING
Views
Sheets
Modify
Regenerate
Switch Dim
Relations
Detail
Interface
Dwg Format
Table
Layer
Set Up
User Attrbt
Symbol
Set Model
Info
Represent

INFO
Mass Props
Names
Measure
Feat Info
Model Info
Comp Info
ParentChild
Layer Info
Drawing

DRAW INFO
Show Draft
Show Ents
Modified Dims
Show Dims
Write Note

Modified Dims

This command is used to highlight *any* dimensions with displayed values that do not match their measured amounts.

Show Dims

This command is used to highlight dimensions based upon the options you specify.

Show Draft

This command is used to highlight all draft entities.

Show Ents

This command is used to highlight drawing entities by layer and type.

Write Note

This command is used to export the contents of a selected note to an external file.

DRAWING ➥ RETRIEVE ➥ INFO ➥ DRAWING ➥ SHOW DIMS

DRAWING
Views
Sheets
Modify
Regenerate
Switch Dim
Relations
Detail
Interface
Dwg Format
Table
Layer
Set Up
User Attrbt
Symbol
Set Model
Info
Represent

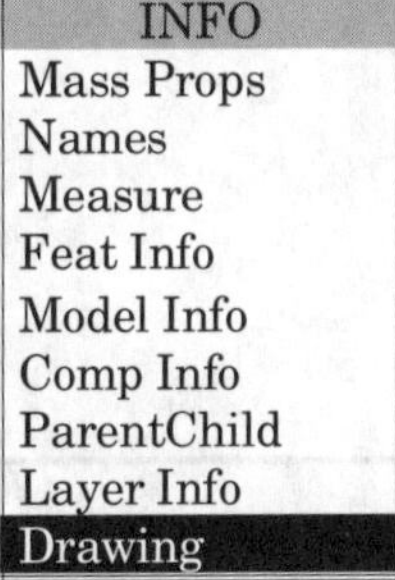

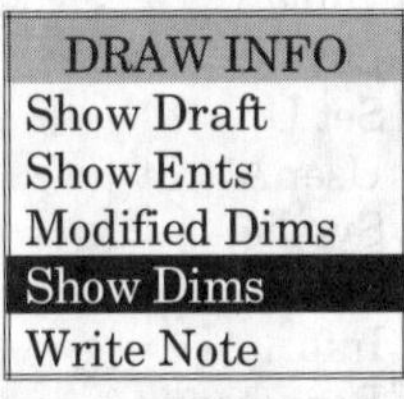

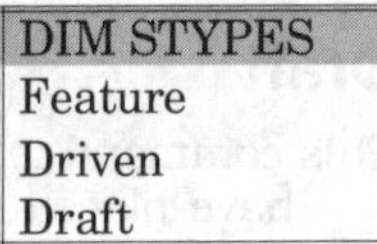

Draft

This command instructs Pro/ENGINEER to highlight the dimensions created in DRAWING mode.

Driven

This command instructs Pro/ENGINEER to highlight the driven and reference dimensions in your drawing file.

Feature

This command instructs Pro/ENGINEER to highlight the dimensions of a model associated to the drawing file.

Regenerate

DRAWING ➥ RETRIEVE ➥ REGENERATE

DRAWING
Views
Sheets
Modify
Regenerate
Switch Dim
Relations
Detail
Interface
Dwg Format
Table
Layer
Set Up
User Attrbt
Symbol
Set Model
Info
Represent

REGENERATE
Model
Draft

Draft

This command is used to regenerate any associative draft dimensions you have placed in your drawing. ***This command does not update the model.***

Model

This command is used to regenerate a model and update the views of that model in the drawing file.

Represent

DRAWING ➡ RETRIEVE ➡ REPRESENT

DRAWING
Views
Sheets
Modify
Regenerate
Switch Dim
Relations
Detail
Interface
Dwg Format
Table
Layer
Set Up
User Attrbt
Symbol
Set Model
Info
Represent

DWG REPRESENT
All Views
Pick Views

Simplify
Restore

All Views

This command instructs Pro/ENGINEER to display the results of the **simplify** or **restore** commands in *all* views.

Pick Views

This command instructs Pro/ENGINEER to display the results of the **Simplify** or **rRstore** commands in the views you specify.

Restore

This command will return a *simplified* view to its original appearance.

Simplify

This command makes complex objects appear in the display with less detail. Most often used to increase productivity when working with complex assemblies.

Setup

DRAWING ➥ RETRIEVE ➥ SET UP

DRAWING
Views
Sheets
Modify
Regenerate
Switch Dim
Relations
Detail
Interface
Dwg Format
Table
Layer
Set Up
User Attrbt
Symbol
Set Model
Info
Represent

DTL SETUP
Create
Retrieve
Modify Val
Save
Quit

Create

This command is used to create a *new* drawing setup file. The parameters stored in the new file will initially be set to the default values.

Modify Val

This command is used to interactively edit the drawing parameters using a text editing window. Once you have modified and saved the desired values, you must perform a **View** ➥ Repaint command to see the results.

Retrieve

This command is used to recall a stored drawing setup file and initialize the drawing parameters contained therein.

Save

This command is used to store the current drawing parameters to an external drawing setup file, which may be recalled and used in as many other drawing files as necessary.

Sheets

DRAWING ➡ RETRIEVE ➡ SHEETS

DRAWING
Views
Sheets
Modify
Regenerate
Switch Dim
Relations
Detail
Interface
Dwg Format
Table
Layer
Set Up
User Attrbt
Symbol
Set Model
Info
Represent

SHEETS
Previous
Next
Set Current
Add
Remove
Reorder
Switch Sheet

Add

This command is used to add a new sheet at the end of a drawing file.

Next

This command is used to move from the current drawing sheet to the next (in ascending numerical order). This command also makes the new sheet the *current* sheet.

Previous

This command is used to move from the current drawing sheet to the next (in descending numerical order). This command also makes the new sheet the *current* sheet.

Remove

This command is used to erase a specified sheet (or sheets) from a multiple-sheet drawing file.

Reorder

This command is used to move an entire sheet from its current order in the numbering sequence to a new location.

Set Current

This command is used to specify which sheet in a multiple-sheet drawing file will be the current sheet by entering the sheet's number via the keyboard.

Switch Sheet

This command is used to move items from one drawing sheet to another. You may move the items to the exact same location on another sheet, or specify a new location using the various options.

DRAWING ➡ RETRIEVE ➡ SHEETS ➡ SWITCH SHEET

DRAWING
Views
Sheets
Modify
Regenerate
Switch Dim
Relations
Detail
Interface
Dwg Format
Table
Layer
Set Up
User Attrbt
Symbol
Set Model
Info
Represent

SHEETS
Previous
Next
Set Current
Add
Remove
Reorder
Switch Sheet

SWITCH OPTS
Switch Items
Switch / Move

Switch Items

This command is used to move items from one sheet to the exact same location on another.

Switch / Move

This command is used to move items from one sheet to a new location on another, using the various options.

DRAWING ➥ RETRIEVE ➥ SHEETS ➥ SWITCH SHEET ➥ SWITCH ITEMS & SWITCH / MOVE

DRAWING
Views
Sheets
Modify
Regenerate
Switch Dim
Relations
Detail
Interface
Dwg Format
Table
Layer
Set Up
User Attrbt
Symbol
Set Model
Info
Represent

SHEETS
Previous
Next
Set Current
Add
Remove
Reorder
Switch Sheet

SWITCH OPTS
Switch Items
Switch / Move

DWG ITEMS
DWG Views
Draft Items
DWG Tables
Done Sel

Dwg Items

This command is used to move selected draft items from one sheet to another.

Dwg Tables

This command is used to move selected drawing tables from one sheet to another.

Dwg Views

This command is used to move selected views of the associated model from one sheet to another.

DRAWING ➡ RETRIEVE ➡ SHEETS ➡ SWITCH SHEET ➡ SWITCH / MOVE

DRAWING
- Views
- Sheets
- Modify
- Regenerate
- Switch Dim
- Relations
- Detail
- Interface
- Dwg Format
- Table
- Layer
- Set Up
- User Attrbt
- Symbol
- Set Model
- Info
- Represent

SHEETS
- Previous
- Next
- Set Current
- Add
- Remove
- Reorder
- Switch Sheet

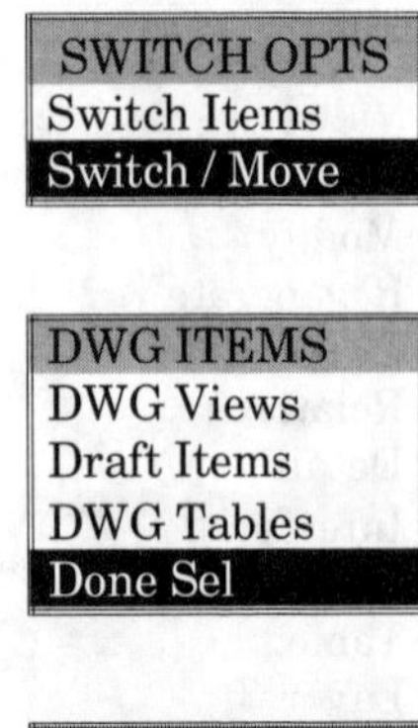

GET VECTOR
- Horiz
- Vert
- Ang / Length
- From - To

Angle / Length

This command instructs Pro/ENGINEER to move the selected type of item from one sheet to another, using a specified angle and distance (from its current location on the existing sheet).

From — To

This command instructs Pro/ENGINEER to move the selected type of item from one sheet to another, using specified points. You will be prompted to select a point to move *from* (on the existing sheet), and then will be automatically switched to the destination sheet where you will define the *to* point.

Horiz

This command instructs Pro/ENGINEER to move the selected type of item from one sheet to another, using a specified distance along the X axis (from its current location on the existing sheet).

Vert

This command instructs Pro/ENGINEER to move the selected type of item from one sheet to another, using a specified distance along the Y axis (from its current location on the existing sheet).

Tables

DRAWING ➥ RETRIEVE ➥ TABLE

DRAWING
Views
Sheets
Modify
Regenerate
Switch Dim
Relations
Detail
Interface
Dwg Format
Table
Layer
Set Up
User Attrbt
Symbol
Set Model
Info
Represent

TABLE
Create
Delete
Move
Enter Text
Copy
Modify Table
Mod Rows / Cols
Repeat Region
Save / Retrieve
Done / Return

Copy

This command may be used to copy a singular *cell* from a drawing table, or the entire table.

Create

This command is used to create a *new* drawing table. You will be prompted to specify the table's justification and if it is to *grow* up or down from the placement point. The default values of **descending** and **rightward** cause the new table to be placed to the right of, and below, your placement point.

Delete

This command may be used to quickly remove an entire table.

Enter Text

This command may be used to place a text string in an empty cell, or to replace any existing text. You will be prompted to identify which cell you would like to edit.

Mod Rows / Cols

This command may be used to insert or remove columns and rows from an existing table. It may also be used to modify the number of characters defining a row or column's size, and to justify the text contained in the cells.

Modify Table

This command may be used to merge table cells together, change the origin of a table, or to turn the display of various table lines on or off.

Move

This command may be used to relocate an existing table from its current position. You will be prompted to select a corner of an existing table to relocate, and to specify the new corner location for the table.

Save / Retrieve

This command may be used to save an existing drawing table as an external file in the current directory. This table file may then be recalled and used in as many other drawing files as necessary.

DRAWING ➥ RETRIEVE ➥ TABLE ➥ COPY

DRAWING
Views
Sheets
Modify
Regenerate
Switch Dim
Relations
Detail
Interface
Dwg Format
Table
Layer
Set Up
User Attrbt
Symbol
Set Model
Info
Represent

TABLE
Create
Delete
Move
Enter Text
Copy
Modify Table
Mod Rows / Cols
Repeat Region
Save / Retrieve
Done / Return

TABLE COPY
Copy Cell
Copy Table

Copy Cell

This command instructs Pro/ENGINEER to copy the contents of one table cell to another.

Copy Table

This command instructs Pro/ENGINEER to copy the entire table to a new location.

DRAWING ➥ RETRIEVE ➥ TABLE ➥ CREATE — Options

DRAWING
Views
Sheets
Modify
Regenerate
Switch Dim
Relations
Detail
Interface
Dwg Format
Table
Layer
Set Up
User Attrbt
Symbol
Set Model
Info
Represent

TABLE
Create
Delete
Move
Enter Text
Copy
Modify Table
Mod Rows / Cols
Repeat Region
Save / Retrieve
Done / Return

TABLE CREATE
Descending
Ascending
Rightward
Leftward
Done
Quit

Ascending

This command sets the first row you create as the *bottom* of the table.

Descending

This command sets the first row you create as the *top* of the table.

Leftward

This command sets the table at right justification (growing to the left of your placement point).

Rightward

This command sets the table at left justification (growing to the right of your placement point).

DRAWING ➥ RETRIEVE ➥ TABLE ➥ MOD ROWS / COLS

DRAWING
Views
Sheets
Modify
Regenerate
Switch Dim
Relations
Detail
Interface
Dwg Format
Table
Layer
Set Up
User Attrbt
Symbol
Set Model
Info
Represent

TABLE
Create
Delete
Move
Enter Text
Copy
Modify Table
Mod Rows / Cols
Repeat Region
Save / Retrieve
Done / Return

ROW / COL OPTS
Insert
Remove
Change Size
Justify
Row
Column

Change Size

This command may be used to modify the number of characters used to define a column's width or a row's height.

Column

This command, when used in conjunction with the **Insert** command, may be used to insert a new column into an existing table. You will be prompted to select the location for the new column, do so by selecting a *vertical* line. If used in conjunction with the **Remove** command, this option instructs Pro/ENGINEER to erase a column of data from an existing table. You will be prompted to select the column to be erased.

Insert

This command may be used to add additional **columns** or **rows**.

Justify

This command may be used **prior** to placing text in any of the table's cells. It is used to specify a text justification (center, left, right) for a selected column of data.

Remove

This command may be used to erase existing **columns** or **rows** from a table.

Row

This command, when used in conjunction with the **Insert** command, may be used to insert a new column into an existing table. You will be prompted to select the location for the new column; do so by selecting a *vertical* line. If used in conjunction with the **remove** command, this option instructs Pro/ENGINEER to erase a column of data from an existing table. You will be prompted to select the column to be erased.

DRAWING ➥ RETRIEVE ➥ TABLE ➥ MODIFY TABLE

DRAWING
Views
Sheets
Modify
Regenerate
Switch Dim
Relations
Detail
Interface
Dwg Format
Table
Layer
Set Up
User Attrbt
Symbol
Set Model
Info
Represent

TABLE
Create
Delete
Move
Enter Text
Copy
Modify Table
Mod Rows / Cols
Repeat Region
Save / Retrieve
Done / Return

TABLE MODIFY
Merge
Remesh
Origin
Line Display

Line Display

This command is used to specify which table lines will be visible and which will not. You will be prompted to select the lines you wish to manipulate.

Origin

This command is used to modify a table's origin. The origin comes into play when inserting new rows or columns. It defines in which direction the table will *grow* as the new items are added.

Merge

This command is used to *join* two or more table cells into one cell definition. The lines separating the cells that were *merged* together will be erased automatically.

Remesh

This command is used to reverse the effects of the **Merge** command.

DRAWING ➥ RETRIEVE ➥ TABLE ➥ MODIFY TABLE ➥ LINE DISPLAY

DRAWING
- Views
- Sheets
- Modify
- Regenerate
- Switch Dim
- Relations
- Detail
- Interface
- Dwg Format
- **Table**
- Layer
- Set Up
- User Attrbt
- Symbol
- Set Model
- Info
- Represent

TABLE
- Create
- Delete
- Move
- Enter Text
- Copy
- **Modify Table**
- Mod Rows / Cols
- Repeat Region
- Save / Retrieve
- Done / Return

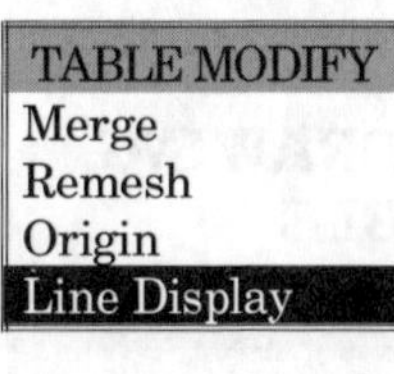

LINE DISPLAY
- Blank
- Unblank
- Unblank All

Blank

This command instructs Pro/ENGINEER to *temporarily* erase any table lines you select.

Unblank

This command instructs Pro/ENGINEER to restore any table lines you have previously blanked. You will be prompted to select the location for each of the blanked lines.

Unblank All

This command instructs Pro/ENGINEER to restore all table lines you have previously blanked.

DRAWING ➥ RETRIEVE ➥ TABLE ➥ SAVE / RETRIEVE

DRAWING
Views
Sheets
Modify
Regenerate
Switch Dim
Relations
Detail
Interface
Dwg Format
Table
Layer
Set Up
User Attrbt
Symbol
Set Model
Info
Represent

TABLE
Create
Delete
Move
Enter Text
Copy
Modify Table
Mod Rows / Cols
Repeat Region
Save / Retrieve
Done / Return

TABLE STORE
Store
Retrieve
List
Store Text

List

This command displays a list of the existing table files, in the current directory.

Retrieve

This command is used to recall one of the stored table files and place the table in the current drawing.

Store

This command is used to save an existing drawing table as an external file in the current directory.

Store Text

This command is used to save only the contents of a table to an external file in the current directory. Each time this command is used a new file will be created using a sequentially numbered file naming convention.

Views

DRAWING ➥ RETRIEVE ➥ VIEWS

DRAWING
Views
Sheets
Modify
Regenerate
Switch Dim
Relations
Detail
Interface
Dwg Format
Table
Layer
Set Up
User Attrbt
Symbol
Set Model
Info
Represent

VIEWS
Add View
Move View
Modify View
Erase View
Resume View
Delete View
Relate View
Disp View
Dwg Models
Represent
Done / Return

Add View

This command is used to add a model view to a drawing file.

Delete View

This command is used to permanently remove a model view from a drawing file. A *parent* view may not be deleted while its children still exist.

Disp Mode

This command displays a sub-menu that allows you to modify the display characteristics of individual views.

Dwg Models

This command is used to choose which model will be the subject of a drawing file view.

Erase View

This command is used to temporarily remove a model view from a drawing file. The view may be recovered using the **Resume View** command.

Modify View

This command displays a sub-menu that allows you to modify many view parameters (arrows, boundaries, origin, orientation, etc).

Move View

This command is used to relocate a view on a drawing file. Any associated views will adjust themselves to maintain alignment.

Represent

This command works with the **Simplify** and **Restore** commands to modify the appearance of a part's features, or of an assembly's components.

Resume View

This command is used to reverse the effects of the **Erase View** command.

DRAWING ➡ RETRIEVE ➡ VIEWS ➡ ADD VIEW

DRAWING	VIEWS	VIEW TYPE
Views	**Add View**	Projection
Sheets	Move View	Auxiliary
Modify	Modify View	General
Regenerate	Erase View	Detailed
Switch Dim	Resume View	Revolved
Relations	Delete View	Full View
Detail	Relate View	Half View
Interface	Disp View	Broken View
Dwg Format	Dwg Models	Partial View
Table	Represent	Section
Layer	Done / Return	No Xsec
Set Up		Of Surface
User Attrbt		Exploded
Symbol		Unexploded
Set Model		Scale
Info		No Scale
Represent		Perspective
		Done
		Quit

Auxiliary

This command is used to create a new view that has been projected in a right angle to a selected surface or along an axis. If using a surface, it must be perpendicular to the screen.

d. FRM.1

Broken View

This command is used to take a very large model, remove a portion of the model (between two specified points) from display and move the remaining portions of the model together. This command should only be used to enhance clarity of an image, and the middle portion selected for removal should not contain important information.

Detailed

This command is used to create a view that contains a portion of the model displayed in another view. Typically, this view is shown at a larger scale factor than the *parent* view to increase visual detail. The orientation of this view type will be determined by the *parent* view.

Exploded

This command instructs Pro/ENGINEER that you do wish the referenced assembly file to appear in an exploded manner. The explode distances are copied from the model file and become part of the view's parameters. They may be modified individually at a later time.

Full View

This command is used to create a view that displays the entire model.

General

This command is used to create a view that is not oriented relative to other views in the drawing file.

Half View

This command is used to remove a portion of a displayed model by a selected cutting plane. The plane can be a planar surface or a datum plane. After selecting the cutting plane, you will be prompted (via reversible arrows) to choose which side of the view you wish to remain displayed.

No Scale

This command instructs Pro/ENGINEER to use the drawing scale as the view's scale.

No Xsec

This command instructs Pro/ENGINEER that you do not wish a cross-section to be displayed in the selected view.

Of Surface

This command is used to create a view that contains only a selected surface of a model. You may not use this command with the **detailed** view type.

Partial View

This command is used to create a view that contains a portion of the complete model that falls within a closed boundary. The parts of the model that fall outside of this boundary will not be displayed in the view.

Perspective

This command is used to create a perspective view of the model.

Projection

This command is used to create a new view that has been projected in an orthogonal direction from an existing view.

Scale

This command is used to apply a scale factor to the model which determines the size of the view created. A scale factor of two will make the view of the model appear at twice its normal size.

Section

This command instructs Pro/ENGINEER that you do wish a cross-section to be displayed in the selected view. This command will only work if the view orientation is such that the existing cross-section plane is parallel to the screen.

Unexploded

This command instructs Pro/ENGINEER that you do not wish the referenced assembly file to appear in an exploded manner.

DRAWING ➥ RETRIEVE ➥ VIEWS ➥ ADD VIEW ➥ SECTION

VIEWS
Add View
Move View
Modify View
Erase View
Resume View
Delete View
Relate View
Disp View
Dwg Models
Represent
Done / Return

VIEW TYPE
Projection
Auxiliary
General
Detailed
Revolved
Full View
Half View
Broken View
Partial View
Section
No Xsec
Of Surface
Exploded
Unexploded
Scale
No Scale
Perspective
Done
Quit

XSEC TYPE
Full
Half
Local
Total Xsec
Area Xsec
Align Xsec
Unfold Xsec
Done
Quit

Align Xsec

This command is used to create a cross-sectional view that is unfolded around a specified axis.

Area Xsec

This command is used to create a cross-sectional view that has only the cross-section area displayed.

Full

This command is used to create a cross-section on a view that covers the entire view.

Half

This command is used to create a cross-section on a view that covers a portion of the model, as defined by a cutting plane. You may reverse the direction of the cutting arrows and thereby reverse which portion of the view is cross-sectioned.

Local

This command is used to create a cross-section on a view that has only a portion of the model sectioned. The portion of the model *inside* the cross-section boundary will be sectioned.

Total Unfold

This command is used to create an unfolded cross-section, displaying the cross-sectioned area and model edges.

Total Xsec

This command is used to create a cross-section that shows the model edges and cross-sectioned area.

Unfold Xsec

This command is used to create an unfolded cross-section, displaying only the cross-sectioned view.

DRAWING ➡ RETRIEVE ➡ VIEWS ➡ DISP VIEW

DRAWING
Views
Sheets
Modify
Regenerate
Switch Dim
Relations
Detail
Interface
Dwg Format
Table
Layer
Set Up
User Attrbt
Symbol
Set Model
Info
Represent

VIEWS
Add View
Move View
Modify View
Erase View
Resume View
Delete View
Relate View
Disp View
Dwg Models
Represent
Done / Return

DISP MODE
View Disp
Edge Disp
Member Disp

Edge Disp

This command is used to control the edge display mode for a selected view.

Member Disp

This command is used to control the display mode for assembly members in a selected view.

View Disp

This command is used to control the display mode for a selected view.

DRAWING ➥ RETRIEVE ➥ VIEWS ➥ DISP VIEW ➥ EDGE DISP

DRAWING
Views
Sheets
Modify
Regenerate
Switch Dim
Relations
Detail
Interface
Dwg Format
Table
Layer
Set Up
User Attrbt
Symbol
Set Model
Info
Represent

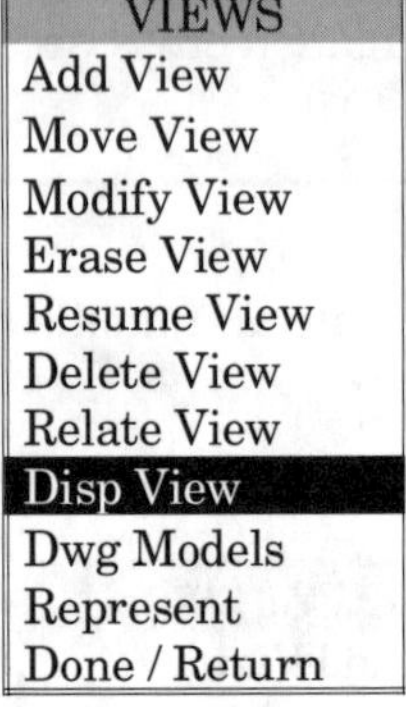

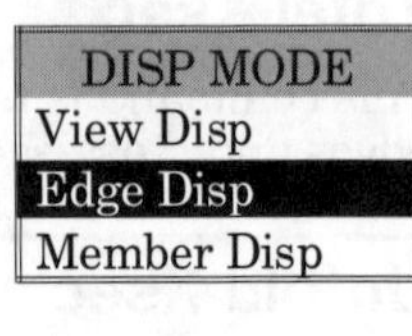

EDGE DISP
Erase Line
Hidden Line
No Hidden
Default
Any View
Pick View
Done
Quit

Any View

This command instructs Pro/ENGINEER to modify the edge display for selected elements in the view to which it belongs.

Default

This command sets the display mode of a selected view to the setting established in the ENVIRONMENT menu.

Erase Line

This command is used to remove a *visible* line from the display.

Hidden Line

This command displays a hidden edge as a hidden line.

No Hidden

This command removes a hidden edge from the display.

Pick View

This command instructs Pro/ENGINEER to modify the edge display for selected elements in the view you specify.

DRAWING ➥ RETRIEVE ➥ VIEWS ➥ DISP VIEW ➥ MEMBER DISP

DRAWING
Views
Sheets
Modify
Regenerate
Switch Dim
Relations
Detail
Interface
Dwg Format
Table
Layer
Set Up
User Attrbt
Symbol
Set Model
Info
Represent

VIEWS
Add View
Move View
Modify View
Erase View
Resume View
Delete View
Relate View
Disp View
Dwg Models
Represent
Done / Return

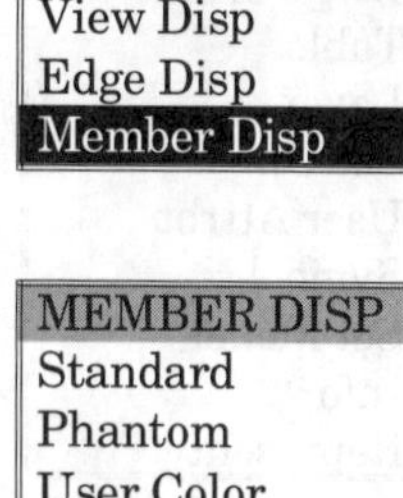

Phantom Font

This command is used to modify the display of assembly members so they appear as phantom lines in the display.

Solid Font

This command is used to modify the display of assembly members so they appear as solid lines in the display.

DRAWING ➥ RETRIEVE ➥ VIEWS ➥ DISP VIEW ➥ VIEW DISP

DRAWING
Views
Sheets
Modify
Regenerate
Switch Dim
Relations
Detail
Interface
Dwg Format
Table
Layer
Set Up
User Attrbt
Symbol
Set Model
Info
Represent

VIEWS
Add View
Move View
Modify View
Erase View
Resume View
Delete View
Relate View
Disp View
Dwg Models
Represent
Done / Return

DISP MODE
View Disp
Edge Disp
Member Disp

VIEW DISP
Wireframe
Hidden Line
No Hidden
Default
Disp Tan
No Disp Tan
Tan Ctrln
Tan Default
From Parent
Det Indep
Done
Quit

Default

This command sets the display mode of a selected view to the setting established in the ENVIRONMENT menu.

Det Indep

This command instructs Pro/ENGINEER that the selected view is independent of it's parent view and will have its own display parameters. This may cause the view to display more slowly.

Disp Tan

This command sets the display mode of a selected view so that *tangent* edges are displayed.

From Parent

This command instructs Pro/ENGINEER that the selected view will receive it's display parameters from its parent view (default).

Hidden Line

This command sets the display mode of a selected view to hidden line.

No Disp Tan

This command sets the display mode of a selected view so that *tangent* edges are not displayed.

No Hidden

This command sets the display mode of a selected view to no hidden line.

Tan Ctrln

This command sets the display mode of a selected view so that *tangent* edges are displayed as centerlines.

Tan Default

This command sets the display mode of a selected view so that *tangent* edges are displayed as specified in the ENVIRONMENT menu.

Wireframe

This command sets the display mode of a selected view to wireframe.

DRAWING ➥ RETRIEVE ➥ VIEWS ➥ MODIFY VIEW

DRAWING
Views
Sheets
Modify
Regenerate
Switch Dim
Relations
Detail
Interface
Dwg Format
Table
Layer
Set Up
User Attrbt
Symbol
Set Model
Info
Represent

VIEWS
Add View
Move View
Modify View
Erase View
Resume View
Delete View
Relate View
Disp View
Dwg Models
Represent
Done / Return

VIEW MODIFY
Ref Point
Boundary
Add Arrows
Del Arrows
Reorient
Mod Expld
Snapshot
Origin

Add Arrows

This command is used to place arrows representing an auxiliary view orientation to its parent.

Boundary

This command is used to modify the cross-sectional boundaries of a selected view.

Del Arrows

This command is used to remove any view arrows placed using the **Add Arrows** command.

Mod Expld

This command is used to specify the distance between cosmetically exploded elements in a view.

Origin

This command is used to modify the origin of a selected view.

Ref Point

This command is used to modify the reference points that define the outer borders of a detailed, partial, or broken view.

Reorient

This command is used to modify a view's orientation after the view has been created.

Snapshot

This command is used to convert the contents of a drawing view into draft elements that are no longer associated to the referenced model.

DRAWING ➥ RETRIEVE ➥ VIEWS ➥ MODIFY VIEW ➥ BOUNDARY

DRAWING
Views
Sheets
Modify
Regenerate
Switch Dim
Relations
Detail
Interface
Dwg Format
Table
Layer
Set Up
User Attrbt
Symbol
Set Model
Info
Represent

VIEWS
Add View
Move View
Modify View
Erase View
Resume View
Delete View
Relate View
Disp View
Dwg Models
Represent
Done / Return

VIEW MODIFY
Ref Point
Boundary
Add Arrows
Del Arrows
Reorient
Mod Expld
Snapshot
Origin

MOD BOUNDAR
View Break
Breakout

Breakout

This command is used in broken views that contain local cross-sections to modify the breakout for the local cross-section.

View Break

This command is used in broken views that contain local cross-sections to modify the view's boundaries.

DRAWING ➥ RETRIEVE ➥ VIEWS ➥ MODIFY VIEW ➥ BOUNDARY ➥ BREAKOUT

DRAWING
Views
Sheets
Modify
Regenerate
Switch Dim
Relations
Detail
Interface
Dwg Format
Table
Layer
Set Up
User Attrbt
Symbol
Set Model
Info
Represent

VIEWS
Add View
Move View
Modify View
Erase View
Resume View
Delete View
Relate View
Disp View
Dwg Models
Represent
Done / Return

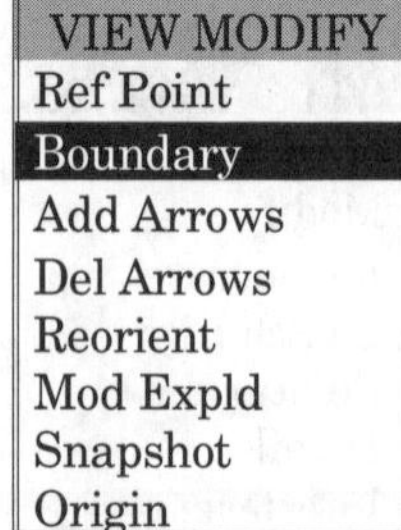

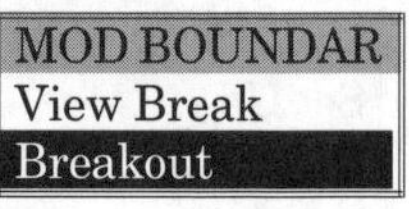

MOD VIEW
Mod Breakout
Add Breakout
Del Breakout
Show Outer
Erase Outer
Done
Quit

Add Breakout

This command is used to insert a breakout definition.

Del Breakout

This command is used to remove a breakout definition.

Erase Outer

This command is used to remove the outer border of a detailed or partial view.

Mod Breakout

This command is used to re-sketch the spline boundary defining a detailed, partial, or broken view.

Show Outer

This command is used to display the outer border of a detailed or partial view.

DRAWING ➡ RETRIEVE ➡ VIEWS ➡ MODIFY VIEW ➡ ORIGIN

DRAWING
Views
Sheets
Modify
Regenerate
Switch Dim
Relations
Detail
Interface
Dwg Format
Table
Layer
Set Up
User Attrbt
Symbol
Set Model
Info
Represent

VIEWS
Add View
Move View
Modify View
Erase View
Resume View
Delete View
Relate View
Disp View
Dwg Models
Represent
Done / Return

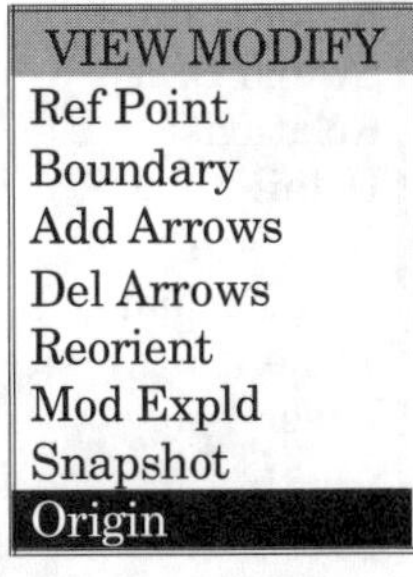

VIEW ORIGIN
On Item
At Center

At Center

This command is used to set the view origin of a selected view to the center of the view's outline (default).

On Item

This command is used to set the view origin of a selected view to an arbitrary location.

DRAWING ➥ RETRIEVE ➥ VIEWS ➥ MODIFY VIEW ➥ REF POINT

DRAWING
Views
Sheets
Modify
Regenerate
Switch Dim
Relations
Detail
Interface
Dwg Format
Table
Layer
Set Up
User Attrbt
Symbol
Set Model
Info
Represent

VIEWS
Add View
Move View
Modify View
Erase View
Resume View
Delete View
Relate View
Disp View
Dwg Models
Represent
Done / Return

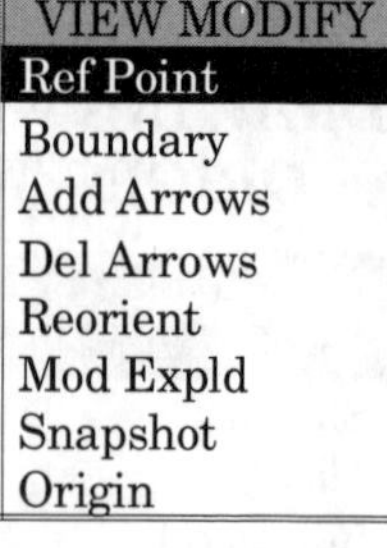

MOD REF PNT
View Break
Breakout

Breakout

This command is used in broken views that contain local cross-sections to modify the reference points of a breakout for the local cross-section.

View Break

This command is used in broken views that contain local cross-sections to modify the reference points of the broken view's boundaries.

DRAWING ➥ RETRIEVE ➥ VIEWS ➥ MODIFY VIEW ➥ SNAPSHOT

DRAWING
Views
Sheets
Modify
Regenerate
Switch Dim
Relations
Detail
Interface
Dwg Format
Table
Layer
Set Up
User Attrbt
Symbol
Set Model
Info
Represent

VIEWS
Add View
Move View
Modify View
Erase View
Resume View
Delete View
Relate View
Disp View
Dwg Models
Represent
Done / Return

VIEW MODIFY
Ref Point
Boundary
Add Arrows
Del Arrows
Reorient
Mod Expld
Snapshot
Origin

SNAPSHOT
All Views
Pick View

All Views

This command is used to select all views for conversion into snapshots.

Pick Views

This command is used to select a single view for conversion into a snapshot.

DRAWING ➡ RETRIEVE ➡ VIEWS ➡ RELATE VIEW

DRAWING
Views
Sheets
Modify
Regenerate
Switch Dim
Relations
Detail
Interface
Dwg Format
Table
Layer
Set Up
User Attrbt
Symbol
Set Model
Info
Represent

VIEWS
Add View
Move View
Modify View
Erase View
Resume View
Delete View
Relate View
Disp View
Dwg Models
Represent
Done / Return

RELATE VIEW
Add Item
Remove Item
Done Sel

Add Items

This command is used to associate detail items to a selected view.

Remove Items

This command breaks the association between detail items and a selected view.

Index

NOTE: All entries in upper case are menu names.

!

A

N

O

P

Q

R

S

T

More OnWord Press Titles

Pro/ENGINEER Books

INSIDE Pro/ENGINEER
Book $49.95 Includes Disk
The Pro/ENGINEER Quick Reference
Book $24.95
The Pro/ENGINEER Exercise Book
Book $39.95 Includes Disk

Mechanical CAD

Manager's Guide to Computer-Aided Engineering
Book $49.95

MicroStation Books

INSIDE MicroStation
Book $29.95 Optional Disk $14.95
INSIDE MicroStation Companion Workbook
Book $34.95 Includes Disk/Redline Drawings/Projects
INSIDE MicroStation Companion Workbook Instructor's Guide
Book $9.95 Includes Disk/Redline Drawings/Projects/Lesson Plans
MicroStation Reference Guide
Book $18.95 Optional Disk $14.95
The MicroStation Productivity Book
Book $39.95 Optional Disk $49.95
MicroStation Bible
Book $49.95 Optional Disk $49.95
Programming With MDL
Book $49.95 Optional Disk $49.95
Programming With User Commands
Book $65.00 Optional Disk $40.00
101 MDL Commands
Book $49.95 Optional Executable Disk $101.00 Optional Source Disks (6) $259.95
101 User Commands
Book $49.95 Optional Disk $101.00
Bill Steinbock's Pocket MDL Programmers Guide
Book $24.95
MicroStation for AutoCAD Users
Book $29.95 Optional Disk $14.95
MicroStation for AutoCAD Users Tablet Menu
Tablet Menu $99.95
MicroStation 4.X Delta Book
Book $19.95
The MicroStation 3D Book
Book $39.95 Optional Disk $39.95
Managing and Networking MicroStation

Book $29.95 Optional Disk $29.95
The MicroStation Database Book
Book $29.95 Optional Disk $29.95
The MicroStation Rendering Book
Book $34.95 Includes Disk
INSIDE I/RAS B
Book $24.95 Includes Disk
The CLIX Workstation User's Guide
Book $34.95 Includes Disk

SunSoft Solaris Series

The SunSoft Solaris 2.* User's Guide
Book $29.95 Includes Disk
SunSoft Solaris 2.* For Managers and Administrators
Book $34.95 Optional Disk $29.95
The SunSoft Solaris 2.* Quick Reference
Book $18.95
Five Steps to SunSoft Solaris 2.*
Book $24.95 Includes Disk
One Minute SunSoft Solaris Manager
Book $14.95
SunSoft Solaris for Windows Users
Book $24.95

The Hewlett Packard HP-UX Series

The HP-UX User's Guide
Book $29.95 Includes Disk
HP-UX For Managers and Administrators
Book $34.95 Optional Disk $29.95
The HP-UX Quick Reference
Book $18.95
Five Steps to HP-UX
Book $24.95 Includes Disk
One Minute HP-UX Manager
Book $14.95
HP-UX for Windows Users
Book $24.95

CAD

INSIDE CADVANCE
Book $34.95 Includes Disk
Using Drafix Windows CAD
Book $34.95 Includes Disk
CAD and the Practice of Architecture: ASG Solutions
Book $39.95 Includes Disk

CAD Management

One Minute CAD Manager
Book $14.95
The CAD Rating Guide
Book $49.00

Geographic Information Systems

The GIS Book
Book $29.95

DTP/CAD Clip Art

1001 DTP/CAD Symbols Clip Art Library: Architectural
Book $29.95
MicroStation
DGN Disk $175.00 Book/Disk $195.00
AutoCAD
DWG Disk $175.00 Book/Disk $195.00
CAD/DTP
DXF Disk $175.00 Book/Disk $225.00
IGES Disk $195.00 Book/Disk $225.00
TIF Disk $195.00 Book/Disk $225.00
EPS Disk $195.00 Book/Disk $225.00
HPGL Disk $195.00 Book/Disk $225.00
CD ROM With All Formats
CD $275.00 Book/CD $295.00

Networking/LANtastic

Fantastic LANtastic
Book $29.95 Includes Disk
The LANtastic Quick Reference
Book $14.95
One Minute Network Manager
Book $14.95

OnWord Press Distribution

End User/Corporate

OnWord Press books are available worldwide to end users and corporate accounts from your local bookseller or computer/software dealer or call 1-800-223-6397 or 505/473-5454.

Wholesale

Domestic Education

OnWord Press books are distributed to the US domestic education market by Delmar Publishers. Call 518/464-3569, Fax 518/464-0301 or write Delmar Publishing Inc. at 3 Columbia Circle, Albany, NY 12203.

Domestic Trade

OnWord Press books are distributed to the US domestic trade by Van Nostrand Reinhold. Call 1-800-842-3636, Fax 606/525-7778 or write Van Nostrand Reinhold at 115 Fifth Avenue, New York, NY 10003.

Europe, Middle East, and Africa

OnWord Press books are distributed in Europe, the Middle East, and Africa by International Thomson Publishing. Call 071-497-1422, Fax

071-497-1426 or write International Thomson Publishing at Berkshire House, 168-173 High Holborn, London WC1V 7AA, United Kingdom.

Asia, Pacific, Hawaii, Puerto Rico, and South America

OnWord Press books are distributed in Asia, the Pacific, Puerto Rico, and South America by Thomson International Publishing. Call 415/598-0784, Fax 415/598-9953 or write Thomson International Publishing at 10 Davis Drive, Belmont, CA 94002.

OnWord Press, 1580 Center Drive, Santa Fe, NM 87505 USA